Whistle Punks & Widow-Makers

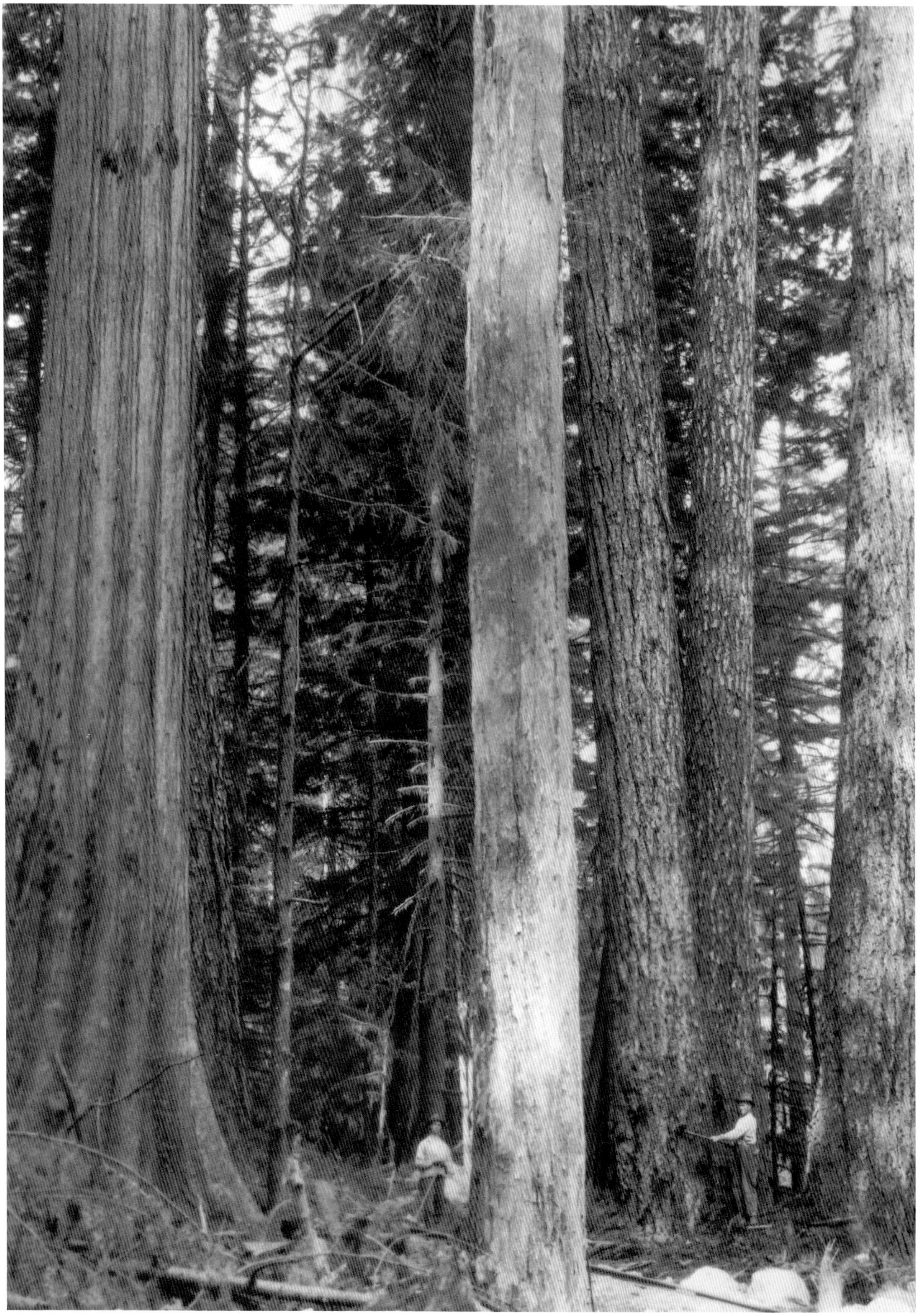

Whistle Punks & Widow-Makers

Robert E. Swanson

With an introduction by Ken Drushka

Harbour Publishing

Copyright © 1993 by Robert E. Swanson

All rights reserved. No part of this book may be reproduced in any form by any means without the written permission of the publisher, except by a reviewer, who may quote brief passages in a review.

Published by
HARBOUR PUBLISHING
P.O. Box 219
Madeira Park, BC Canada V0N 2H0

Published with the assistance of the Canada Council and the Government of British Columbia, Cultural Services Branch

Edited by Ken Drushka
Page design and composition by Fiona MacGregor
Cover design by Roger Handling
Cover painting by Sheila Gibbons (courtesy of the Truck Loggers' Association and the British Columbia Forest Museum), photograph by Bob Cain
Author photograph by Stephen Osborne

Frontispiece: Handfallers at Merrill & Ring's Theodosia Arm operation in 1926. (VPL 1546)

Photograph credits: BCARS – British Columbia Archives and Records Service; CDM – Courtenay & District Museum & Archives; CRM – Campbell River Museum and Archives; CVA – City of Vancouver Archives; IWA – International Woodworkers of America, Local 1-80; UBC – University of British Columbia Library, Special Collections; VPL – Vancouver Public Library.

Canadian Cataloguing in Publication Data

Swanson, Robert E.
    Whistle punks & widow-makers

ISBN 1-55017-090-2
    1. Logging – British Columbia – Anecdotes.
2. Logging – British Columbia – History. I.
Title.
SD538.3.C2893 1993   634.9'8'091109   C93-091648-4

Printed in Canada

CONTENTS

Introduction 7
Eight-Day Wilson 14
Bull Sling Bill 18
Saul Reamy 28
Curly Hutton 37
Charlie Anderson, the Lucky Swede 42
P.B. Anderson, the Swede Who Didn't Go Home 48
John Blacklock "Daddy" Lamb 55
Reverend George Pringle 59
The Horseplay Murder 64
Len Cary and his Famous Six-Spot 67
Bill Barbrick 72
"Broomhandle" Charlie Snowden 80
Harry Smith, the Timber Cruiser 85
The Handlogger's Christmas 90
Locies Lived and Breathed for Inspector Jack Short 94
Chris Meyland, the Bull of the Woods 100
Big Jack Milligan 104
Jessie James, the Big Time Logger 108
Seattle Red 112
Matt Hemmingsen 118
George Hanney and the Flying Dutchman 125
Red Morrison 129
Harry Todd, King of Donkey Punchers 134
Fred Gallant, the Bull Cook 139
Old Hickory Palmer 144
Henry Norman, the Skidder Rigger from Louisiana 150
Index 159

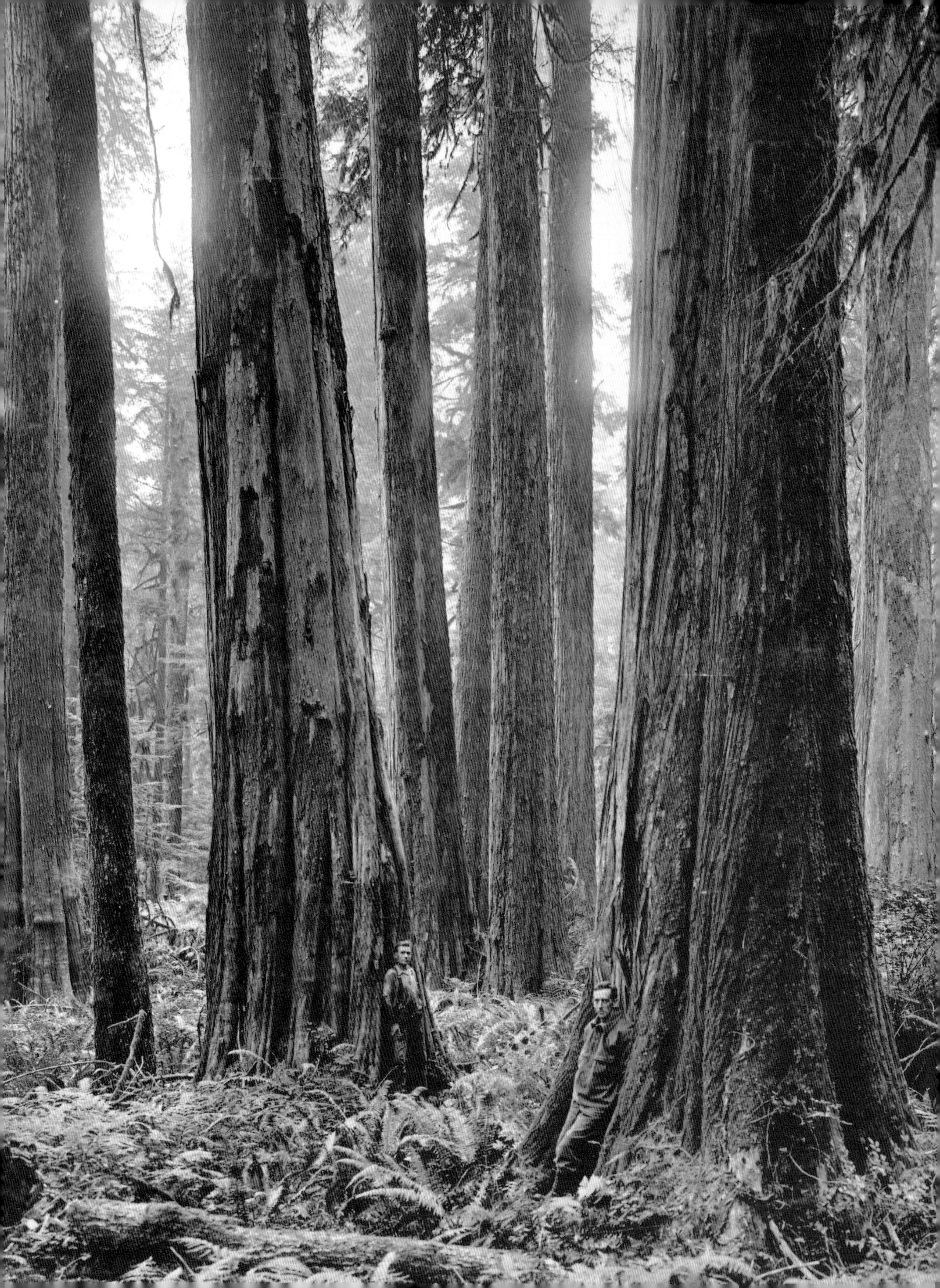

INTRODUCTION

Bob Swanson's stories are the stuff from which legends grow. With his poetry, they constitute a major portion of coastal British Columbia's logging lore from the steam era between the two world wars. Of the tens of thousands of people who worked in and around the West Coast logging industry during this period, only a small handful left a record of what they encountered. Swanson towers above his contemporary scribes, not only for the volume of material he produced, but for the skill with which he wrote and for what it tells us.

This collection provides a nice balance to the recent new edition of Swanson's poetry, *Rhymes of a Western Logger*. But there is a major difference between that work and these stories. The poetry is about the mechanics of logging, and about logging as a way of life. *Rhymes of a Western Logger* celebrates a certain kind of existence. The stories in this edition are about individual loggers, or in some cases composites, who for one reason or another were well known within the logging fraternity and beyond. These are the near-mythical, sometimes larger-than-life characters—the Heroes of Logging, if you like.

For the most part these characters were not the founders of great corporate dynasties—the H.R. MacMillans, Bob Filbergs, Prentice Bloedels or Henry Mackins. They were the hookers, fallers, bull cooks and locomotive engineers—the working stiffs who from sheer force of personality rose above their workaday callings and became the subject of interminable bunkhouse bullshit sessions and barroom storytelling. They are the real-life Paul Bunyans of the West Coast.

Swanson has said that he wrote his poetry in an attempt to do for loggers what Robert Service did for the Klondike miners. What prompted him to undertake this task when he did—the late 1930s and early '40s—was the realization that an era was passing and that without a poet to record it, the meaning of a way of life would die.

"Without Homer, the Greeks would amount to bugger all," is Swanson's succinct way of putting it. Herein lies a clue to the nature and character of Swanson himself. These stories record the doings of an amazing array of remarkable characters. Yet they reveal very little of, arguably, the most memorable character of the whole

Opposite: Timber cruisers in a stand of cedar owned by the Capilano Timber Company on the north shore of Burrard Inlet, 1918. (UBC Capilano Timber Co. Collection, Album 33)

period of West Coast history between 1920 and the present — Swanson himself.

He admits to being born in 1905. This may or may not be true; long ago he learned the world is populated by large numbers of narrow-minded scissorbills who, as a matter of course, dismiss anyone over the age of sixty as irrelevant old fossils. Having accomplished more since he reached the put-out-to-pasture age than the rest of us do in a lifetime, Swanson learned quickly the value of discretion and the virtues of lying about his age.

He may have got into the habit as a young child. Born in England, Bob moved to East Wellington (near Nanaimo) with his family when he was but a wee lad. At the age of three he talked himself into a ride on the little mining locomotive that ran past his house. He spent his pre- and early-teen years tagging along with his father, a mason who rebricked boilers in the mines and sawmills that flourished around Nanaimo in the years just after World War One. Slack times were spent carving tombstones in a back room of the Globe Hotel.

By the time he was fifteen (by his accounting), and with a few summers' experience punking whistle on local logging shows, Swanson got the job of his youthful dreams — firing a steam donkey. He learned fast and began a lifelong drive to acquire the certifications required to perform the work that intrigued him, steam engineering. He built his first steam whistle at about the age of twenty, for the Nanaimo Lumber Company. At one stage he fast-talked his way into a job as leverman on a duplex loader, one of the most exacting occupations in the woods.

In the late '30s he decided he wanted to be a boiler inspector for the government. "It was the only work that provided a pension," he says today, and adds with a gleeful chuckle, "fully indexed." The position called for a university degree, which he proceeded to acquire from the University of British Columbia without ever attending as a regular student.

This job sent Swanson into practically every logging camp and mill operation on the coast, and put him in close contact with the people he describes in this collection. He spent long hours on the Union boats going into remote camps, riding the busses to Port Alberni or Campbell River, in the company of talkative working loggers. During the war he spent much of his time working at the Aero Timber operation in the Queen Charlottes, charged with maintaining the steam equipment with which workers brought out the spruce used for fighter planes and bombers.

It was during this period that he began his career as a poet. It started with a weekly talk show on radio station CJOR, a somewhat unusual occupation for a provincial civil servant. It was an outlet for his enormous energy and enthusiasm, which were at times probably not overly appreciated in the bureaucracy. "Government people are the absolute shits to do anything with," Swanson says. "They have no goddamn brains. They think out their rear ends."

After publishing four small books of poetry, which sold eighty-some thousand copies, he set about writing these stories. They appeared in *Forest and Mill*, a semi-monthly newspaper published by forest industry associations which contained a variety of articles of interest to workers in the logging and sawmilling industries. The material for some of them was obtained in his travels; for others he deliberately sought out and interviewed the people he wrote about, many of them living in obscure retirement by the time he found them. This aspect of Swanson, the thoroughness with which he undertakes any task, is what sets him apart from the other scribblers of his day. He knew his undertaking was important – to record the culture of a people. As he says, "You do the best you can."

I came to appreciate that statement, and the man for whom it is a rule of life, when his poems were reprinted in 1992. By coincidence I had a book on logging published at about the same time by the same company. The publisher suggested we undertake a joint book promotion trip, which became known as the Great Logging History Tour.

We spent more than a week together, in every major community on Vancouver Island, up Howe Sound and out the Fraser Valley. We signed vast numbers of books in every bookstore we could find, appeared on stage, radio and television. It was a gruelling schedule – organized, we suspected at the time, by an adolescent brimming with the energy of the sexually ungratified.

Once he realized I was a competent driver and he was not required to perform that function, Swanson sat back and talked. We were passing through and stopping in places where he had spent more than seventy years working, living and writing. He knows every little nook and cranny of these regions, knew the people who lived in scores of the older houses, worked in camps and mills long since gone and overgrown or replaced with subdivisions and strip malls.

Wherever we went, throngs of people appeared. I quickly discovered most of them were there to see Swanson, to get his autograph on a tattered copy of *Bunkhouse Ballads* or *Rhymes of a Haywire Hooker* that "my John, who passed away in '68, bought on the Union boat in 1942." In public readings the front rows invariably contained a few logging widows, elderly ladies and veterans of camp life who sat with tears streaming down their faces while Swanson reeled off such poems as "The Dying Logger's Lament" or "The Ballad of the Shanty Queen."

And then on to the next gig, with Swanson discoursing energetically in the seat beside me. I had expected to have to tuck him into bed early and spend my evenings alone in a hotel room, reading peacefully over a glass or two of Scotch. Fat chance. Swanson was good until midnight or more, matching me drink for drink and

Victoria Lumber & Manufacturing Company's skidder #3 at Fanny Bay in 1942. This machine was rebuilt in the Chemainus shop by master mechanic Bob Swanson in the 1930s. Loading engineer, Jack Mahony, greasing in front of donkey; head loader, Angus McKay, on flatcar; skidder engineer, Tom Couzins, standing between boilers. (BCARS 80587)

regaling me with stories of his ancient and recent past. What I had thought would be a nostalgic junket by a couple of old farts who wrote about logging in the good old days, was turning into a Behan-like odyssey through the hinterlands of the West Coast.

I learned that Swanson has a lightning wit and a ribald sense of humour. His quick response to a very nice-looking woman's invitation to examine the locally preserved steam locie—"We'd love to have you"—was a libidinous grin and a gallant: "My dear, there's nothing I'd love more than to have you in an old locomotive." In Duncan, he dealt with an obnoxious heckler in a few sharp words, and later explained that years of dealing with deputy ministers had taught him how to handle such types. And he told me the story of Harold Bronson, one-time superintendent at Bloedel, Stewart & Welch's Franklin River Camp. One evening he accompanied Bronson on a trip to Port Alberni's Goat Ranch—for the sake of literary research, he insisted—and while Bronson was off entertaining one of the ladies, Swanson interviewed some other occupants of the house. He wanted to know what they knew about Bronson. "Bob," they assured him, "you haven't lived until you've slept with Bronson." That, for Swanson, is an accolade worth more than the honours heaped upon the captains of industry he has also met.

The tour was also occasion for me to discover the wide range of Swanson's working interests. For example, a quick, unannounced visit to John Casanave's Western Challenger truck plant revealed that Swanson had pioneered the development of air brakes on logging trucks. In fact, there is probably no aspect of transportation in BC that does not, in some way, bear the stamp of his inventive mind. The development of railways, roads and bridges, tramways, balloon logging, ski lifts and myriad other industrial and public pieces of the infrastructure all came under his domain in his various roles—which included Chief Inspector of Railways, head of mechanical engineering on the PGE Railway and a host of other positions.

Heading north out of Duncan one rainy afternoon to catch a ferry, Bob directed me to turn off and drive up the Nanaimo River valley. Far up the valley we pulled into a clearing in the forest where a strange-looking structure stood. It was Swanson's whistle farm. Here he developed and tested the air horns he designed and built for use on diesel-powered locomotives and ships, after the demise of steam. The Rube Goldberg setup included the boiler off an old steam donkey, diesel compressors, and enough gauges, piping, horns and other apparatus to equip a fair-sized factory. This is where the horns that play *O Canada* to downtown Vancouver every noon were tested. Here, he is in his element, starting up engines, opening and closing valves, puttering proudly as he explains the workings. Today the air horns are a large and active business that occupies him several days a week. As we leave and race for the ferry, he brings down vile curses on the heads of the vandals who for years have plagued him here.

Just when I think we have a shot at the one o'clock sailing, he directs me off the road again, this time into the outer suburbs of Nanaimo—East Wellington country. This is where he grew up, and he remembers every hill and dale, explaining where the mining and logging railways ran and who lived where, and pointing out the tree under which he kissed a girl for the first time. I followed his directions through a back route of dubious appearance and we made the ferry—where, to his great delight, he found a piece of raisin pie which he had unsuccessfully sought in every restaurant we had entered during the previous week.

When I dropped him off at his Shaughnessy home, Bob was as fresh and enthusiastic as the day we left. I, decades younger than him, was exhausted. The next morning he phoned me. "Is there anyplace else they want us to go? How about Abbotsford? Or Squamish? And Powell River. A fellow has to do the best he can, you know." Then he chuckled to himself on the other end of the line. "That's all a steer can do."

The stories in this book, written almost fifty years ago, are full of the same boisterous enthusiasm that invigorates Swanson today. These were his kind of people. Given a choice, they were the people he spent his evenings talking with, listening to and writing down their stories, and in the process becoming the greatest living storehouse of coastal logging history. But don't read these stories as literal history. Recording the factual, non-fictional account of events was never Swanson's intent. He was more interested in capturing the mythic or legendary aspect of his subjects. He is a poet, not an historian, and as such felt quite comfortable enhancing historical accuracy to heighten a mythic nuance.

This is not to say he was careless with his details. Far from it. What makes these stories just as readable today as they were half a century ago is the authenticity of the detail. When Swede Oleson chased Lea Curtis into Camp O with an axe, it wasn't just any old axe. It was a double-bitted axe. When he mentions a line on a donkey he'll usually tell you its diameter to within an eighth of an inch. What endears him to his readers is that he has been there and knows whereof he speaks.

What Swanson captures best of all are subtleties of wit and humour that define the character of working loggers. Distinguished by its own peculiar vocabulary, logging talk of the kind written here is one of those branches of language used and esteemed by working people the world over. It employs faint twists of meaning and subtle syntax to make a statement that goes right over the heads of the bosses, tycoons and other lesser mortals who inadvertently make their way onto the claim. The language itself is part of the myth. And the myth, the legends and logging lore are an essential part of the culture of the Pacific Northwest.

It is fortunate that these stories are being published at this time. In the past few years most collections of *Forest and Mill* have been destroyed by librarians and archivists unaware of the significance of their content to our cultural history. It is good to know that our great-grandchildren will be able to find comfort and meaning in the lives of people like Seattle Red, Eight-Day Wilson and Jesse James. Because of Bob Swanson, the loggers will live forever.

– KEN DRUSHKA

Victoria Lumber & Manufacturing Company's #3 skidder rigged to a 178-foot spar tree at Copper Canyon in 1946. (Mauno Pelto Collection)

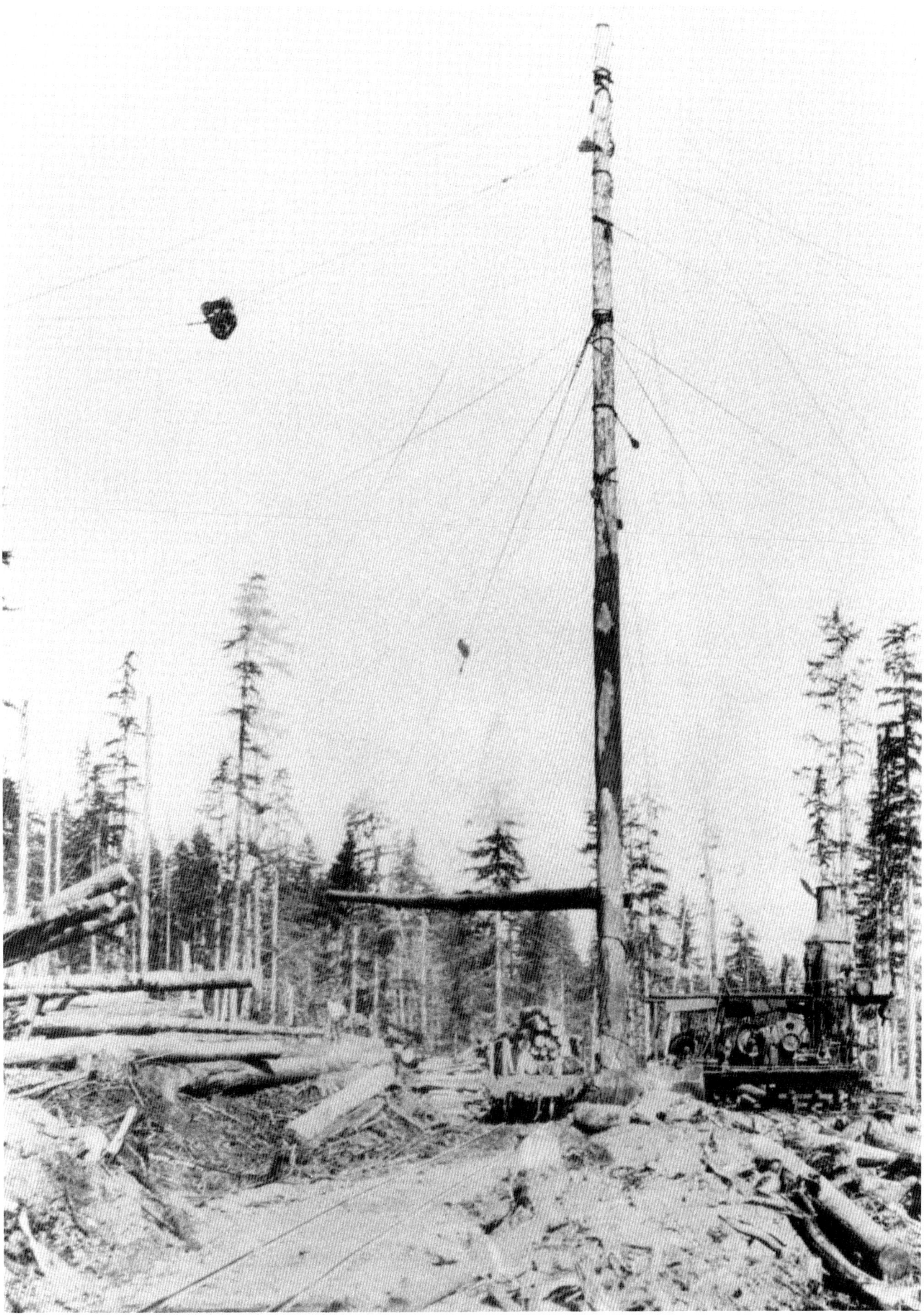

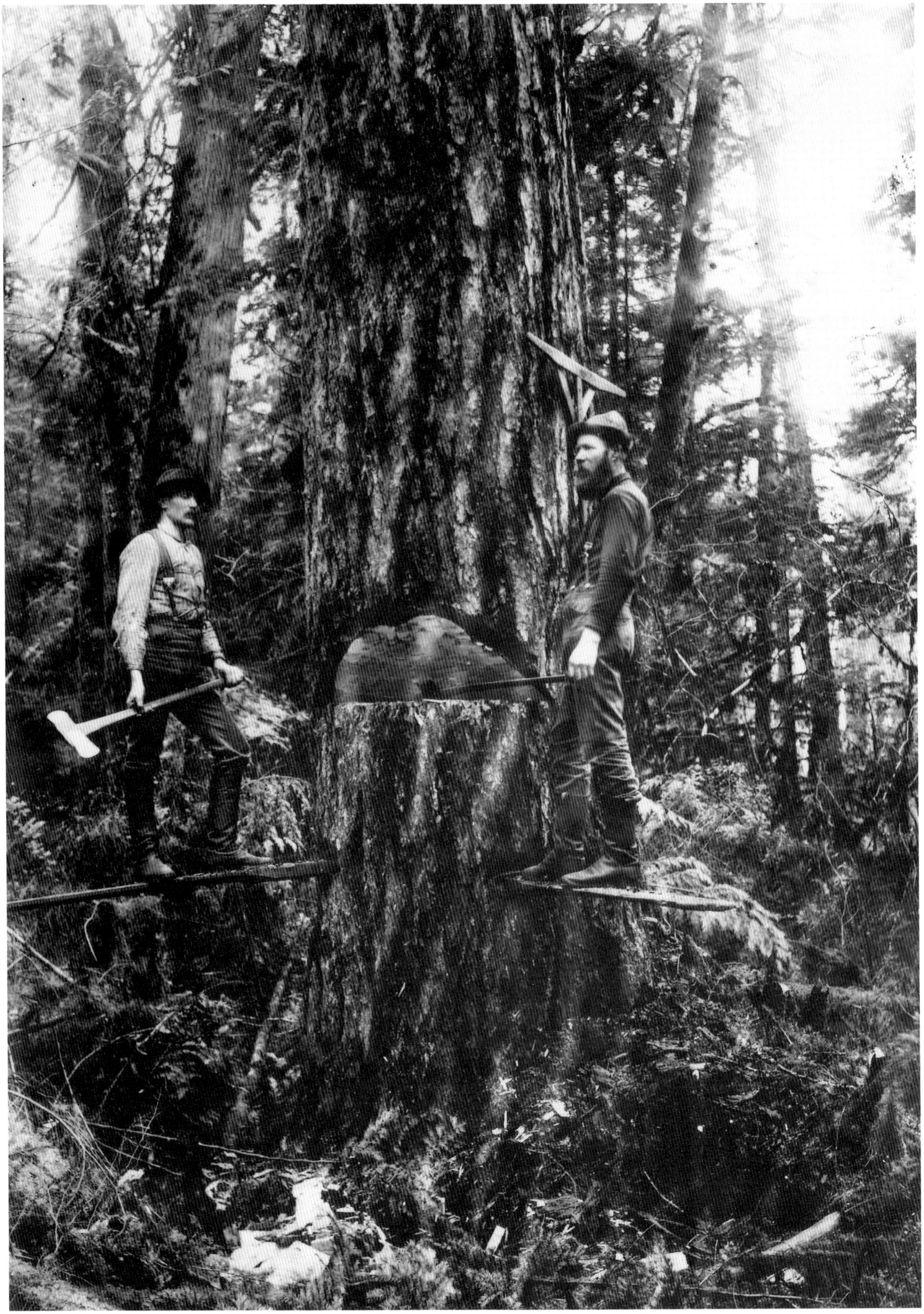

EIGHT-DAY WILSON

Since time immemorial loggers have been famous short stakers. At the building of King Solomon's temple the loggers in the forest of Lebanon are said to have quit at the slightest provocation, jumped on their asses and journeyed in to the nearest town to go on a wing-dinger. This trait in loggers has been evident right down to recent times. During the Roaring Twenties, "Seattle Red" was reputed to have worked for seventy-nine logging companies in the state of Washington in one year.

In the BC woods the character who seems to have been the Short Staker Cup Holder was a logger known as "Eight-Day Wilson." His Christian name at the present writing seems to be unknown. Even Bill Black told me that "Old Eight-Day" just happened, and even though Black hired him out for twenty years his card read "Wilson, Eight-Day: Country of birth – unknown."

Eight-Day Wilson roamed the big clearing back in the days when they cut the stumps above the swell – ten or twelve feet high – and left the big blue butts to rot in the woods. Those were the days when a logger carried his own roll of blankets and hit town wearing caulked boots, stagged-off pants and a mackinaw shirt.

He got the name of Eight-Day for obvious reasons – he was a fair-weather eight-day staker. After eight days on the job in any camp he would usually bunch her and head back to town. He was known to hire to a camp and after sampling the grub and washing his socks, beat it back to the bright lights and go on a tear.

A favourite stunt of Eight-Day's was to hire to a camp as hook-tender and lie in camp the first day to sober up. After getting a mug-up from the cook in the middle of the morning he would saunter up the skid road and poke around a little. Being a boomer he was sure to know most of the boys in camp from the PF man down to the whistle punk, from whom he would get the lowdown on what the new outfit had to offer in the way of a logging show. If the ground was rough or he didn't like the donkey puncher he would wander back to camp, roll his blankets and head back to town.

One time at Loughborough Inlet, where Jack Phelps was pushing camp, Eight-Day pulled the pin because a green whistle punk sat in his place at the table – that was excuse enough for even a home-guard to tell the ink slinger to write-her-out but it didn't take much

Opposite: Handfalling with axes before the use of crosscut saws, in Vancouver about 1890. (Bailey Bros. photo, CVA TR P10 N24)

Oxen skidding logs from Quilchena to railway in Kitsilano, about 1890. (CVA Log P12 N8)

of an excuse for Old Eight-Day. Once when a flunky handed him cold hotcakes he bunched her right then and there.

At Cowichan Lake his favourite stunt boomeranged on him. Eight-Day Wilson had hired out for Matt Hemmingsen and had completed the customary eight days. He needed an excuse to quit, so he resorted to the infallible test: to throw a caulked boot up in the air. If it stayed up, he stayed in camp for another eight days. The trick had never been known to fail. He even had his duffel bag packed in readiness when he threw up the caulked boot and it went through the stovepipe hole in the bunkhouse and stayed up. To save face Wilson had to stay another eight days, but he went on a record bender when he hit the bright lights with a sixteen-day stake.

"Eight-Day Wilson" was the last of the great short stakers. His kind are dying out and loggers are gradually losing the wanderlust from their nature and beginning to settle down. Eight-Day Wilson, ace short staker, embarked on his farewell journey to the camps of the Holy Ghost some years ago. He said goodbye to the bright lights and then walked off the end of the Ballantyne Pier in Vancouver, BC.

Bunkhouses like this one did not encourage loggers to stay put for long periods of time. (CRM 13312)

The dining room at Capilano Timber Company's Camp A in 1918. (UBC Capilano Timber Co. Collection, Album 33)

BULL SLING BILL

According to the legends of Paul Bunyan, the key man at the Great Onion River Camp was a bunkhouse bard by the name of Truthful Thomas whose sole duty was to keep the bunkhouse cranks amused by telling them yarns and far-fetched stories. And according to the legends on the Queen Charlotte Islands, A.P. Allison employed a bunkhouse bard by the name of Bull Sling Bill to keep the boys happy up in spruce country.

Bull Sling Bill was a typical logger of the old school and had done every job in the woods from bull skinning to bull cooking. Many of his escapades as a rip-snorting logger will be handed down from father to son through generations of loggers, but the bedtime stories he's told from Wisconsin to Alaska will someday belong to the folklore of the timbered West.

To hear Bill tell it, he was born back of a stump in the woods of Michigan, but I've found out he was actually born in the Sand Hills of Wisconsin in 1874 and that his real name was William Strausman. He was only fourteen when he started in the woods back there driving the ice sleigh; that was a sleigh with a water tank on it to sprinkle the roads so they'd freeze over. It was a cold job. Bill once told me he had on seven undershirts and the wild hay was growing right out the back of his collar. By the time he'd done everything in the Wisconsin woods from river driving to kicking the town saloon to pieces, he came west to Aberdeen, Washington. That was in 1892 when bull teams were in their heyday, and Bull Sling Bill could sling bull with the best of 'em. He was PF man at one camp down there where the grub wasn't very good. It was a small camp that had a lady cook, and she was curious to know what the letters PF stood for, so she asked Bill at the supper table one night. He told her that PF stood for Practically Famished, and after that the grub improved.

Most men at some time in their lives have had the urge to go to the South Seas, and Bull Sling Bill was no exception. So in 1900 he shipped out to the Philippine Islands to log hardwood. Bill said to some people it might have been hardwood but to him it was all hard work. Most of the help he had to work with were dark-skinned natives and a white man loomed up like a white leghorn hen in a flock of black Minorcas. One time they were white and the only

Falling spruce for airplane construction during World War One in the Queen Charlotte Islands. (UBC BC 1456/62)

good thing about it, according to Bill, was they were all the same colour before quitting time. After that he came back to BC, where he worked as a hooktender on the old ground lead shows.

Along around 1910 old Bill got tired of roaming around and thought he'd settle down in one camp for a change. He'd always heard the Hastings outfit at Rock Bay was a home-guard camp, so he hired out for Saul Reamy and landed up there with his blankets. He hung his clothes on a nail but next morning they were lying on the floor and someone else's clothes were on the nail. An old home-guard with whiskers down to his knees told him in no uncertain terms that *that was his nail*, Bill would have to drive in one of his own. So Bill bunched her at the Rock and headed for the Charlottes.

Massett Timber Company's mill and camp under construction at Buckley Bay in the Queen Charlotte Islands, 1918. (UBC BC 1452/62)

It was on the Charlottes in 1915 that Bill Strausman took his logging seriously and became a camp foreman for the big Buckley Bay outfit. Skagway Jack was pushing Camp 8 and Bull Sling Bill was the big noise over at Camp 9, on Masset Inlet. It was during the first World War and men were hard to get. Finally they hired a crew of Haida Indians. Bill tried to make a rigging crew out of them, but the turn was hung up most of the time and the crew was properly stumped. Bill went out in the brush to fight a few hang-ups and show them how it should be done. He was a great man to swear, Bull Sling Bill, and the Haida crew thought sure he was placing some kind of curse or other on them. At last one day he pulled the original hooktender's oath. That is, he got down on his prayer bones and in a loud voice called upon God to come down from heaven — not to send His son, it was a man-sized job, but to come himself in a birchbark canoe and paddle this bow-and-arrow crew clear off the claim. Bill opened his eyes, which were still fixed on the heavens, and saw one of his Haidas thirty feet up a big sapling and still climbing. He hollered to him and asked what the big idea was and the man answered that if God was coming down, then he was taking no chances — he was climbing halfway up to meet him.

Another time on the same job, old Bill got into a scrap with a big Swede faller. He'd been running around the Swede and slugging him for half an hour, but the Swede just wouldn't fall down. Finally old Bull Sling said: "Holy old mackinaw! Do I have to put the wedges to you and holler *Timber-r-r-r* before you'll topple over?"

When the Buckley Bay outfit folded up in 1924, old Bill came

Bucking airplane spruce, Queen Charlotte Islands, 1918. (UBC BC 1456/62)

Opposite: "Selective logging" 1918-style. The valuable spruce was removed and the unwanted hemlock and balsam were left behind. (UBC BC 1456/62)

down to Vancouver and went on a big drunk. It was such a wing-dinger that he was broke flat in a few days. But being a good talker he talked his way so far into debt that he went back to the Charlottes and built a shack at Queen Charlotte City, where he hibernated until his debts were outlawed in town. This shack was his headquarters for the next twenty years, and from there he hired out as a hooktender, dropped back to rigging slinger, chokerman, whistle punk and was bull cooking when I met him. While we were talking one day in the wood yard, a big black raven perched itself on a snag nearby and uttered those guttural noises for which ravens are famous. Thinking of Edgar Allan Poe and his immortal raven, I remarked to Bill that I thought it would be easy to teach a raven to talk, to which Bill replied: "They can talk already. Listen to that

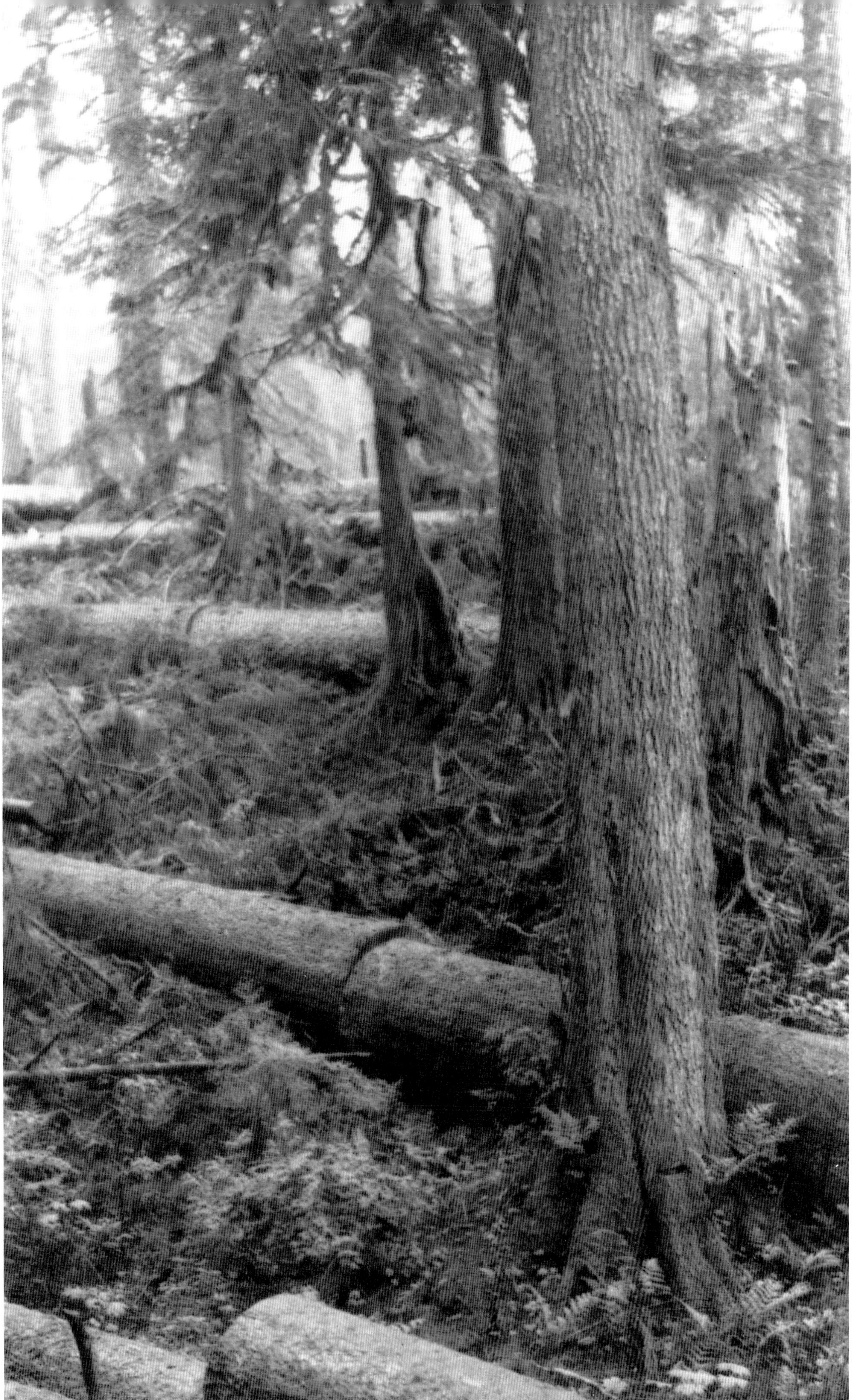

A 10-foot diameter spruce log, Queen Charlotte Islands, 1918. (UBC BC 1456/62)

Opposite: A log chute in Queen Charlotte Islands, 1918. The logs were probably skidded down the chute with a steam donkey located on the beach. (UBC BC 1456/62)

black devil up there teasing me now. Can't you hear it saying 'Yer broke – yer broke – yer broke'? That one's been following me around for the last twenty years telling me that."

It was in the bull cook's shack at nights when the boys used to gather round that old Bill was at his best. He was telling us a yarn one night about the time he worked in Aberdeen, Washington, when I butted in and remarked that it must have been a big outfit he was working for.

"No," said Bill, giving me a dirty look, "it was a small one, they only ran fourteen sides." So I kept quiet and listened.

Allison had just built a new pre-fabricated cookhouse and we were saying what a dandy it was. "'Tain't nothing," said Bill. "You shoulda seen the one I had down in Grays Harbour . . . steam potato mashers! The flunkies were on roller skates whisking in and out around the tables. The dining room was so big the head flunky used

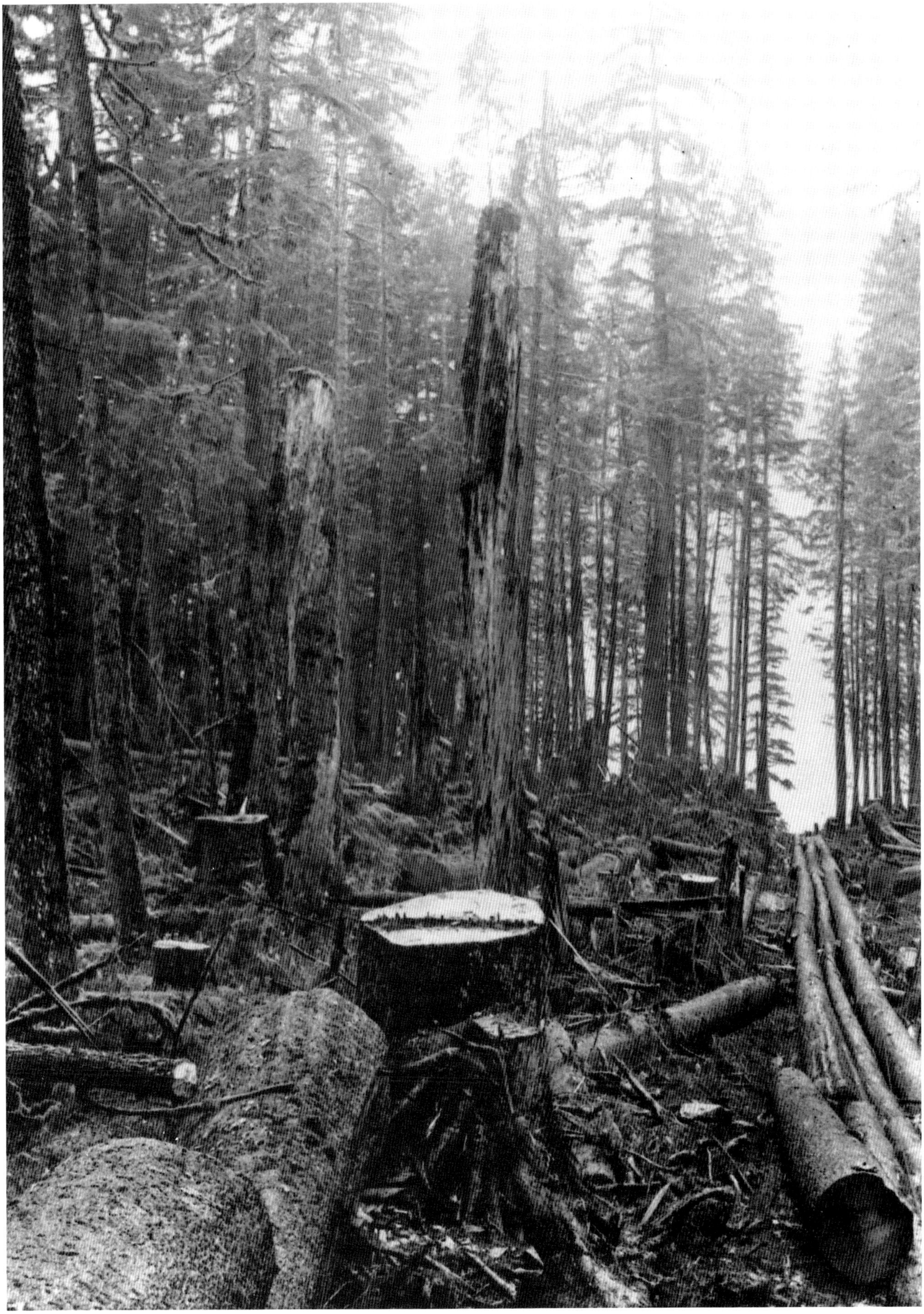

Massett Timber Company's Buckley Bay sawmill operating, 1918. (VPL 3876)

to stand at the dish-up table and direct traffic with a pair of binoculars. You guys think this is a big cookhouse, why you ain't seen nothing."

Perhaps some of Bill's most far-fetched tales were about the time he was in the Klondike in '98. "It was so cold," said old Bill, "that the smoke used to freeze off the top of the stovepipe in big chunks, and in the spring I'd thaw out the smoke for mosquito smudge." One night in Dawson, Bill said he went broke and in the morning only had $2.00 cash in his pocket. As breakfast cost at least $2.50, Bill thought he'd have a whirl at the roulette wheel, so he put $2.00 on red and it paid off $4.00. He kept on winning double-or-nothing until by nine he was $20,000 ahead. After breakfast he still had seven or eight hundred dollars small change in his pocket, so he went back

Log booms and Davis rafts, Massett Timber Company, 1918. (UBC BC 1456/62)

to the magic wheel. But it must have been in reverse, because by eleven o'clock he'd lost $20,000 and $10,000 of his own money. Such were the tales of Bull Sling Bill as told to the loggers on the Queen Charlotte Islands.

Bill Strausman, the famous bunkhouse bard of the Queen Charlotte Islands, passed on to the land of the heavenly timber at Queen Charlotte City in 1944, and was buried in a rock grave alongside Boxcar Pete on the Charlottes he knew so well. And with his passing, the logging fraternity lost one of the most colourful and big-hearted characters that ever roamed the BC woods, a man who was known to have actually given the shirt off his back to one of his fellow men with only one stipulation, and that was he had to be a twenty-one-jewel guy.

SAUL REAMY

Logging crew with ox team near Vancouver, about 1880. (VPL 19767)

I'm going to tell you about a pioneer logger who let daylight into the swamps along False Creek – right here in Vancouver. That was back in the days before the shrill blast of a donkey whistle shattered the slumbering silence of the timbered valleys and the primitive ox teams plodded along the skid roads that are today Hastings and Powell Streets; and the beacon that guided the sailing ships into Vancouver's harbour was the burner at Hastings Sawmill. The vast wealth of the new timber industry was beginning to be realized when the foundations of our great city were rising from the settlement around Hastings Sawmill – then known as Gastown –

Handfalling with axes in Kitsilano, about 1885. (CVA TR P37 N26)

when Saul Reamy, the famous foreman logger of the Hastings Company, arrived on the scene back in 1885.

Saul Reamy was born in New Brunswick in 1860. Although he'd driven horses and oxen back in that Maritime province where they logged in their own primitive way, the skid road system and the size of the BC timber presented many new problems. But it wasn't long before the merits of Saul Reamy, the foreman at Hastings Camp, were being discussed by the loggers in Gassy Jack's Saloon.

It was in 1899, after the townsite of Vancouver was pretty well logged off, that Saul moved his outfit up to the immortal Rock Bay where he started in to log with railroad as well as horses and bull teams, and it was around the turn of the century that steam donkeys began to appear in the woods. These first little steam pots had a vertical engine driving a gypsy drum, but later the single drum job appeared with a horizontal engine and a line horse was employed to pull the rigging back to the woods. Saul Reamy soon mastered the art of logging with steam and the faithful line horse. He had one

Hastings Sawmill, at the foot of Dunlevy Avenue, 1888. (CVA MI P21 N57)

View down the jack ladder at Hastings Sawmill, 1898. (CVA MI P53 N50)

Looking north at Hastings Sawmill, 1913. The original mill burned in 1898 and was rebuilt. (CVA MI P60 N67)

Hastings Sawmill, 1919. (VPL 3937)

Moffat's skid road between Sechelt and Gibsons, 1903. (CVA OUT P878 N427)

John Hudson, wood bucker, on Gambier Island, 1906. (Phillip Timms photo, VPL 4995)

line horse they called "Gerry" that had caulks in its shoes, could walk a log, and even chewed tobacco.

In 1907 Saul Reamy opened up Camp O on East Thurlow Island for the Hastings Company, where he put the little locomotive they called "Curly" and a few single drum donkeys to work. Along around 1909 the manager, a man by the name of Sandy McNair, sent up a brand-new two-drum donkey on a float along with two reels of line, and the night it arrived in camp Saul Reamy ranted and raved about the stupidity of those dudes in town shipping up a donkey with two drums and a reel of ⅝-inch line. Whoever heard of a donkey with two drums? Next morning Reamy had the crew drag out the mainline—by hand—and they unloaded the new machine and moved her two miles up the skid road, leaving the haulback in camp because they didn't know what it was for. They'd been yarding with her for three weeks using a line horse when the big shot arrived in camp and Saul lit into him. "Why that other drum's the haulback, Saul," said the manager. "Guess I'd better ship you up a hooktender that's worked with them newfangled rigs—

Building a skid road bridge at Hastings Sawmill camp on East Thurlow Island, 1890s. (BCARS 14288)

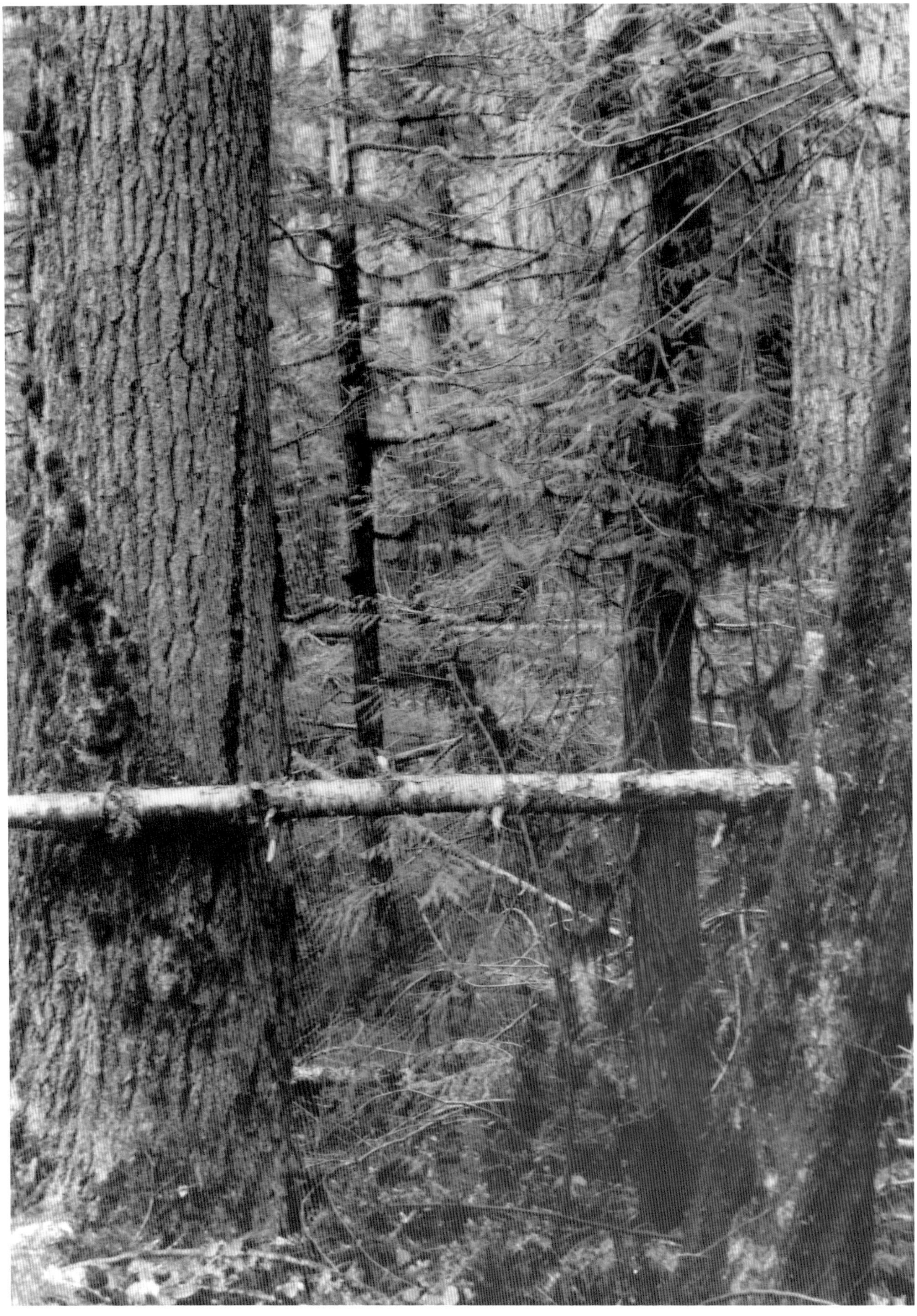

Fallers working at BC Mills Timber & Trading's Rock Bay Camp, 1919. (Leonard Frank photo, VPL 3928)

some of the camps are even using geared locomotives."

"Don't send one of them up here," said Reamy. "We've too many gear wheels in camp now."

Reamy always ran a good camp and his men usually stayed with him for years. Many of them had never worked for any other outfit. Greasy Bob Graham and Whiskey Shannon were cooks up at Camp O on East Thurlow Island, famous in their day, and many are the tales told about them. One thing Saul Reamy hated to see was a man quit the outfit. One time the black hooktender, Lea Curtis, had a run-in with a Swede sniper by the name of Sam Oleson, and the Swede chased Lea right into camp with a double-bitted axe. Lea got dressed up in his town clothes and was heading for the dock when Saul Reamy met him and said surely he wasn't leaving, to which the hooker replied: "It's this way, Boss—far better to say there he goes than here he lies." And the hooker caught the boat. Another time, after Saul had been raised to the sublime position of superintendent, a new foreman up at Rock Bay issued orders that as several of the old-timers had outlived their usefulness, they should be laid off and sent to town. When this reached Saul Reamy's ears he walked up and asked the new foreman if there were plenty of seats left at the table, and if these old-timers were still able to enjoy a hearty meal? The foreman said yes, they could still eat plenty.

Opposite: Stand of BC Mills Timber & Trading Company timber at Rock Bay, about 1919. (Leonard Frank photo, VPL 3929)

"Well," replied Reamy, "those are the men that made the Hastings Outfit and as long as they can still eat, there'll always be a job for them at Rock Bay."

Saul Reamy was man-catcher for the Hastings Company in Vancouver in 1921, trying to hire fallers who could do their own saw-filing and had their own blankets. It was on this job that he befriended an old section man who'd worked for him for years. And, like the prophet of old who cast his bread on the waters and in many days it returned to him, a few years later the section man died and left Saul Reamy $20,000 in his will.

Thus after the Hastings Company liquidated in 1925 Saul was enabled to live in comfort for the few remaining years of his life, until he died in the Vancouver General Hospital in 1928 at the age of 68 years. And by 1960 the little locomotive that Saul had used on Thurlow Island—the one they called "Curly"—could be seen in Vancouver's Hastings Park, erected as a monument to the early pioneer loggers like Saul Reamy, who founded one of BC's biggest and most important industries.

CURLY HUTTON

The old Climax stalled a hundred yards below the top of the hill on the 8 percent grade. They'd have to wait for steam but that wasn't what was burning Curly Hutton up. He'd told old Daddy Lamb a thousand times to keep those pigs of his off the track and today was the last straw—he'd widened on her and ran over six of them—all prize pigs, too. Now Daddy Lamb had docked six pigs at $35 each from his paycheque, added to the $12.50 he'd had to pay for yielding to the temptation of guzzling Mulligan Mag's barrel of beer the week before. It looked like a pretty sober Christmas was in store for the old hogger. So they would just have to wait for steam often from now on!

That was railroading at Daddy Lamb's big home-guard camp back in 1921, and the locie engineer was none other than the famous Curly Hutton, ace character of the BC logging railroads.

Curly Hutton was born June 3, 1889 in Ottawa, Ontario where he attended Sunday school and church. He came to BC in 1906 and spent the first four years learning to be a machinist at Schakke's Machine Works in New Westminster. In 1910 he joined the CPR as wiper and worked up to firing on the mainline between Vancouver and Field.

Once a man gets railroading into his blood he can never leave it alone, so in 1921 he went up before Jack Short and got his locie ticket. This made him a full-fledged throttle jerker on any man's logging railroad, and he promptly hired out to Daddy Lamb's at Menzies Bay. After the episode with the pigs, Curly wondered if he should have hired out as a PF man.

It was at Nimpkish Lake where Curly Hutton had the old 3-Spot go through a bridge under him when he and his fireman jumped clear on the opposite side and hit the ground the same time as the locie did. Norman English, the manager, came out to see the wreck and said to Curly, "Lad, you sure are lucky—there must have been a hole in that bridge."

"Well, if there wasn't," answered the curly-headed one, "there sure is a big hole there now!"

Rule "G" held no terrors for Curly Hutton, so long as an inspector didn't catch him. One time up at APL, on the old China Creek Line, it was quite apparent Curly and his crew were in good spirits—Scots

Capilano Timber Company's 70-ton Climax locomotive, 1918. (UBC Capilano Timber Co. Collection, Album 33)

spirits at that! Everyone wondered where the bottle had come from so early in the day.

Those were the days of the famous "Goat Ranch" at Alberni, and the bull bucker had been celebrating in port. On his way back to camp he had called in at the Goat Ranch to pay his respects to the virtuous landlady, intending to catch the locie when she whistled by on the last trip back. They treated the bull bucker so well that he lost track of time, and it was dark when he finally did leave to walk ten miles up the railroad to camp. On the way back, there were two long trestles, one a very high one and the other a long low one. Being well-oiled for the journey, the bull bucker fell between the ties and hung on by his fingers for dear life, thinking for sure that he was on the high trestle! When he let go he expected to plummet into the ravine, but to his surprise he fell only a few inches and a full bottle of Scotch rolled out of his pocket which he didn't miss until he hit camp. On hearing this at breakfast Curly winked at his head brakeman, and a wink is as good as a nod to a thirsty railroader. They stopped at the low bridge first trip to the beach and drank a toast to the Goat Ranch.

Curly Hutton always was proud of his locie even though it was only a Climax, which breed of locie Curly always seemed to have the doubtful fortune to run. At Squamish Curly's luck changed

A Bloedel, Stewart & Welch Shay at Myrtle Point, the company's first camp in BC, in 1926. (VPL 1510)

when he went to work for George Moore, for at that camp he leaned out the cab window of a 50-ton Shay.

On the first Sunday he decided to wash out the boiler and in so doing found there to be a lot of scale in her. "We'll fix that," said Curly, and he sent his fireman to the cookhouse for a few sacks of potatoes while he took off the dome cover. Together they dumped three sacks of potatoes into the boiler and bolted up the dome cover. Next morning Curly's Shay hooked onto the crummy and the pops lifted. Foam blew from the pops and the cylinder cocks until the locie looked like an iced Christmas cake. The crew piled out of the crummy like sheep out of a stock car. Old George Moore blew his top, but Curly told him she only needed to be blown down.

"Don't do it here," bellowed Moore. "She might blow up!"

When Curly did finally get her blown down and working again the white stuff from the boiler had whitewashed one side of fifteen

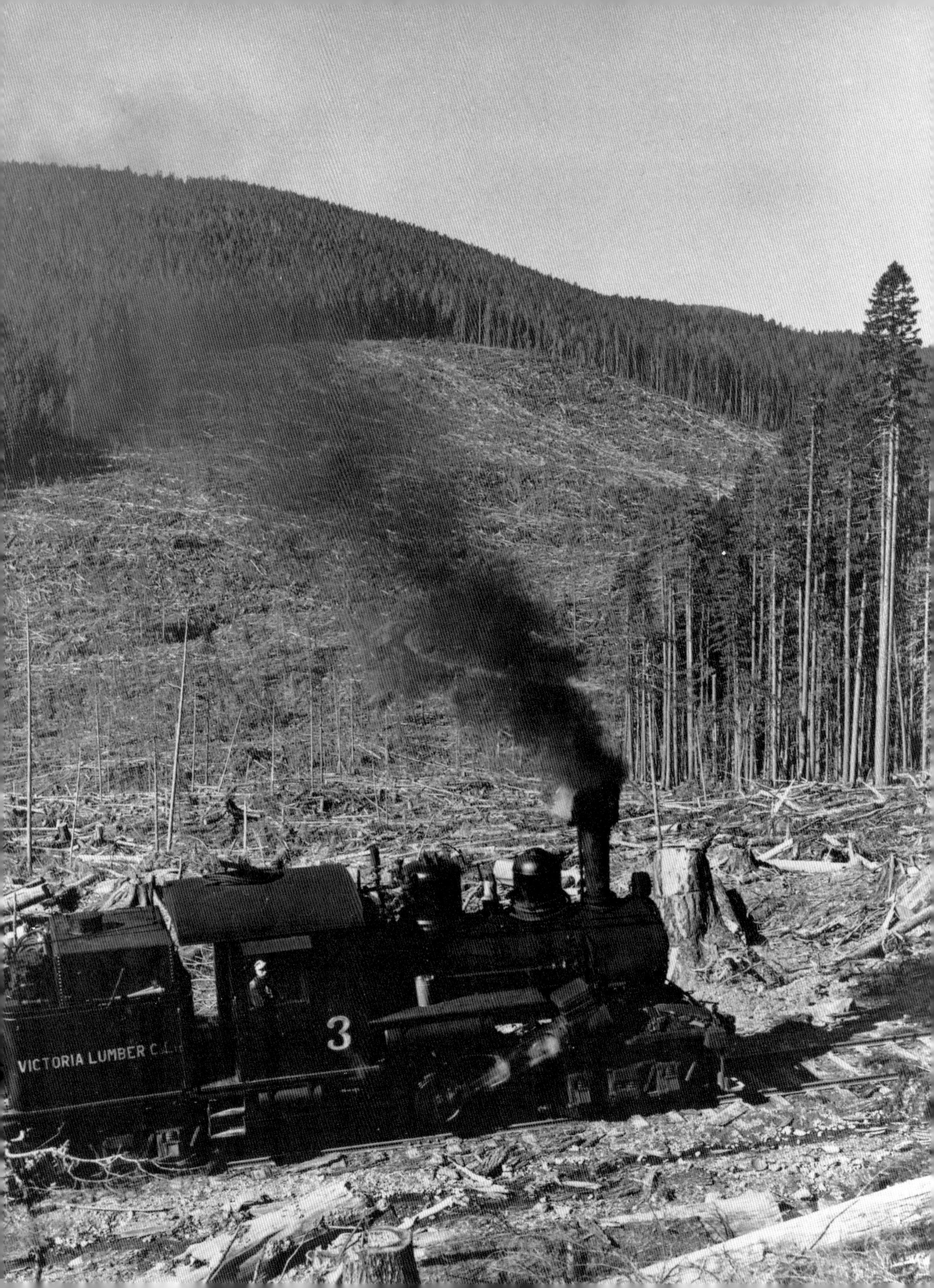
3
VICTORIA LUMBER C

PGE box cars. But Curly was proud. The inside of that boiler was as clean as a silver dollar even though on the outside she looked like a white elephant.

By 1940 Curly Hutton was getting fed up with uphill brakemen and downhill firemen. His mind went back to his yesteryears at Capilano Timber Company, Orford Bay, Pitt Lake, and many other outfits now gone by the board. He remembered, too, the time he punched road donkey and tightlined the skid road pig clear off the skid road, dumping Charlie Peter the PF man into the salal brush. He longed for a change. Being a machinist, among other things, Curly joined the Royal Canadian Navy in September 1940, and it was no time until he was Chief ERA on HMS *Magedonia*, chasing submarines on the Atlantic. It was then he longed to be back in the woods punching an old Climax.

Once when docking at Halifax, Curly got the bell for "full speed astern." Something went wrong and his ship went clear through the dock. He was called on the mat about it and asked to explain what happened. He told the Skipper that when he got the bell he threw the Johnston bar into reverse and opened the back sanders, but as usual the back sanders weren't working and she must have gone into a skid!

When the war was over the famous old seagoing locie engineer went back to his Climaxes, and his many friends were glad to know that in 1946 he was still in overalls leaning out the right-side cab windows of the No. 102 Climax up at Englewood.

I met him in town that Christmas and asked him what he thought of diesel locies, to which he replied: "Look, yer leanin' back in the cab watchin' the old air gauge while yer brakeman sets up the bugs. Ya hear yer pump workin'—umpah-um-pah, like she's pantin' for the journey that lies ahead of ya. Ya glance over at yer fireman and he hollers, *highball!*, ya pull the old whistle cord and put the air in release. Brother, that's living. When they bring them newfangled diesels into the woods I'm going back into King George's Navy."

Victoria Lumber Company's Climax No. 3, about 1940. (Jack Cash photo, UBC BC 1930/549/97VL 4021)

CHARLIE ANDERSON THE LUCKY SWEDE

Charlie Anderson, otherwise known as the "Lucky Swede," sat in the door of his log cabin at Vananda on Texada Island, and watched the summer sun sink like a ball of fire behind the mountains that fringe the Gulf of Georgia. That sun seemed to rekindle in his aging memory the experiences of his forty-eight years on the Pacific Coast. How well he remembered his recent years in the BC woods and his fishing trips up and down the coast with its winding inlets! They reminded him of his native Sweden where he was born in 1856—but he felt he was nearing the end of his journey and it wouldn't be long before that old sun would be setting without him to watch it.

Charlie Anderson was world-famous in his day, for he was the Lucky Swede who'd made and spent a million dollars before his luck went back on him.

It was forty below the night that Charlie Anderson first mushed into the town of Dawson with his sleigh loaded down with gold. There was plenty more where that came from, too, for he'd struck it rich on Eldorado Creek. He tilted a heavy poke of gold on the bar, and it was drinks on the Lucky Swede. The saloon seethed with excitement, the Lucky Swede was instantly a favourite with the Belles of the Yukon as he amused himself by dropping gold nuggets down the backs of their necks. That was in January of 1897; by the next winter the big gold rush of 1898 was in full swing.

Anderson had landed in New York from Sweden in 1887, and came west to Tacoma where he worked in the woods and heard stories of gold up in the Yukon. In 1893 he and eight other men went north to Alaska by boat and tramped overland through the Chilkoot Pass. When they reached the Yukon River they made a boat and sailed 400 miles downstream, where they pitched camp and went on a moose hunt. It was on the hunt that Anderson met a "Squawman" who said that he and his Indians had been fishing salmon on Rabbit Creek and that one of the Indians called "Skookum Jim" had seen nuggets glistening in the river bed. When the Anderson party reached the creek they found it all staked and being worked, for this creek turned out to be Bonanza. Even Eldorado Creek had been staked to its source, and it seemed like they'd arrived too late.

About this time, three other men came into camp, named Wilfred Olar, Al Parks and Al Fair. They had staked claim number 29 on the Eldorado, but wanted to sell it. Now Anderson had $7000 in gold cached in his cabin on Glacier Creek, but he couldn't get it until the rivers froze, so he talked the manager of Alaska Commercial Company into loaning him $800 with which he bought a quit-claim deed from the three men — and claim 29 on the Eldorado was all his. When the rivers froze, he got his money from the cabin on Glacier Creek, paid his debt and headed out for his claim on the Eldorado.

It was Christmas Eve of 1896 when he reached Eldorado. Within a month his prospecting proved the ground to be fabulously rich, and it was then that he headed into Dawson with his sled loaded with gold nuggets while the mercury was dipping down around 40 below. So the piano played and so did the Lucky Swede! What did

Lumber ships loading at Brunette Sawmill, New Westminster, 1895. (CVA MI P59 N45)

The Brunette Sawmill in 1917. (VPL 9659)

it matter if he blew the works? There was plenty more on Eldorado. He hired a crew and paid the highest wages ever to be regularly paid in the North country, and by 1905 he had over a million dollars.

Now when a man has a million dollars there's some mighty tall tales told about him, especially if he's a Swede in the Klondike. It is said one of the dance hall belles fell madly in love with him. Maybe she did. Anyhow she had competition, for the others felt themselves falling for the Lucky Swede too. Such a storm of protest was raised that the women decided to play a game of poker with the Lucky Swede as the stakes. The Swede sat and watched the game to see there was no dealing from the bottom of the pack, and he is said to have taken the winner to Frisco where they were married. He is also supposed to have given her her weight in gold as a wedding present and bought and furnished a fancy home in Seattle.

But the woman left him, and the Lucky Swede decided to be sensible and invest his money in real estate in Frisco. It was then that his luck deserted him, for the San Francisco earthquake and fire in April of 1906 destroyed in a day the fortune he had taken fourteen years to make on Eldorado Creek, and he was left flat broke.

In 1907 Charlie Anderson landed back in Vancouver and took a job in the old Brunette sawmill. Later he went to Texada Island where he worked in the woods and the lime quarries, and put in the odd season fishing up and down the coast. But always he was going to go back to the Klondike where he was sure he would again strike it rich as he had in '97.

Charles John Anderson, the world famous Lucky Swede, died in the Powell River Hospital in 1936. With his passing the sun set on one of BC's most colourful and respected pioneer citizens. In the words of the loggers' minister, the Reverend George C.F. Pringle: "He was always industrious, always cheerful; his habits such as would make me, with a clear conscience, recommend him for an eldership in one of our churches."

The kitchen of the 120-man Powell River Company camp at Kingcome Inlet, 1915. (UBC BC 1930/276)

Powell River Company office. This was the most modern camp on the coast at this time. (UBC BC 1930/276)

Bunkhouses at the Powell River Company camp in Kingcome Inlet. (UBC BC 1930/276)

A foreman's room. Crewmen's rooms slept four, and were steam-heated and equipped with kerosene lamps and desks. (UBC BC 1930/276)

P.B. ANDERSON THE SWEDE WHO DIDN'T GO HOME

An Arctic wind stabbed at his face as the big Swede thumbed the only four-bit coin he had in the world. The Northern lights flashed across the bowl of the sky and poured back to the pole with the gentle flow of milk as he gazed into a frosted restaurant window in Dawson. How was he going to get a meal for fifty cents when the cheapest meal in that frostbitten town was a dollar and a half? The Sourdoughs thought he was crazy to even bother towing twenty tamarack poles down the river to Dawson, but when he'd sold them all his debts were paid. Fifty cents was all that was left, but he could look the world in the face for he owed no man.

As he stepped into a nearby saloon two men were arguing about setting up a whipsaw to saw boat lumber. Pete strode over and settled the argument by telling them straight he was not only a logger but a millman as well—but right now he was hungry. What about eating? In the morning he'd set up the whipsaw for them.

To Pete Anderson the smell of the newly sawn lumber the next morning was nothing new, and hammering a kinked whipsaw into shape on the top of a stump was as easy as falling off a log, in fact easier, for he was no ordinary logger. He was P.B. Anderson, and inside of a week he was a partner in the little outfit cutting boat lumber for the gold-crazed prospectors.

That was back in 1897 and P.B. Anderson was determined to make good. He'd lost every cent he had in a sawmill deal in Bellingham and had come North to start all over again. His past experience should pay him dividends in this golden land of opportunity for he was a man who could do things with his hands and wasn't lazy.

Born in Onsjo, Sweden, P.B. Anderson arrived in the Swede's Paradise—Minnesota—in 1885, and took a job on a farm. The Chicago, Burlington and Northwestern Railroad was under construction so the young Swede quit hoeing corn and went to work for the railroad. When he'd saved $400 he decided to go back to Sweden. He walked two hundred miles into St. Paul, where ostensibly he was going to get his teeth fixed, but where his $400 dwindled down to seven bucks. So Pete went back to farming again. This was to be the turning point in his life, for the farmer was a

P.B. Anderson's first sawmill, at Cedarville, Washington, in 1893. (UBC P.B. Anderson Collection)

Scotsman. The green Swede couldn't speak English and knew nothing about business ethics, and it was the canny Scotsman who taught him both. He learned to speak English by reading a chapter from the Bible out loud every night.

It was in the fall of '86 that Anderson went to work in the woods of Wisconsin for the Weyerhaeuser Timber Company. They were driving the river and P.B. started sacking logs. He was in the water most of the time and about the only log he could ride at first was a "schoolmarm" – but his partner, a little French Canadian, he was so good he could ride a cork! Well, one night, along toward dark, Pete fell in and his partner decided they'd ride a schoolmarm across the river to camp, but when they got there they'd moved the camp and had to tramp ten miles over a tote road up river looking for it. But that was logging in those days.

Two years later Anderson came out West and got a job in the Northern Pacific roundhouse in Tacoma and when he'd saved $45 decided to go into business for himself. He landed in Bellingham, called Whatcom in those days with a population of 400 people. He took a contract clearing land at $50 an acre, but there wasn't much money in that so he bought in on a grocery business and inside of two months wore out two wheelbarrows making deliveries. This business he sold for $1000 and went homesteading, and from that he got into the butcher business; but when he went into the real estate business he lost every dime he'd made and had to go back bookkeeping at seventy-five bucks a month to pay his debts.

One night when P.B. was working late on the books a fellow by the name of Bill Peters came around, and knowing that Anderson was a go-getter, offered him a partnership in a small shingle mill on the Nooksack River. Anderson accepted the offer and they logged with bull teams and sold the shingles at a dollar and five cents a thousand. They went broke but the enterprising Swede managed to pay his debts — and it was a hard winter, was '96 — and by the spring he had only seven dollars when he headed for the Klondike by signing on as the ship's butcher and second cook to work his passage.

The Klondike was a tough country but Anderson was a tough man, and with his knowledge of sawmilling, logging and boatbuilding he landed back in Seattle a few years later with a real stake — a pot full of money. Now some men would have settled down and taken it easy, for that stake was real money in those days, but not P.B. Anderson. He built a big sawmill at Blanchard, Washington, bought a steam donkey and ten horses, and went in for logging in a big way. He'd logged and milled about fifteen million when the market went haywire and the outfit went belly-up, and P.B. was left holding the bag for $34,000.

He didn't sit down and curse his luck but borrowed $250 to build a mill out of round timbers in the Columbia Valley, which he operated until he could finally pay his debts. Then he moved to BC, where he started all over again.

Anderson's first outfit in BC was on Flat Island, and when that was finished he took a contract to log for Hastings Mill, and in 1916 he took his two boys, Dewey and Clay, and started again on his own up at Knox Bay and then Grassy Bay. It was at Grassy Bay that Stewart Holbrook was timekeeper, and many of BC's leading superintendents and managers got their start in life with old P.B. Anderson up the BC coast.

By 1929 it looked like P.B. Anderson was at last a tycoon and had made a success of the logging business. He had an interest in a big sawmill at Marpole, too. But the slump of 1929 broke him again, so he was back where he had started thirty years before. Now the old fellow had always paid his debts, so when he and his sons went into partnership to start all over again up at Green Point Rapids, the banks and the wholesale house gave them the backing they needed. By 1937 they were back on their feet and going stronger than ever.

Opposite: P.B. Anderson cruising timber on West Thurlow Island, 1917. (UBC P.B. Anderson Collection)

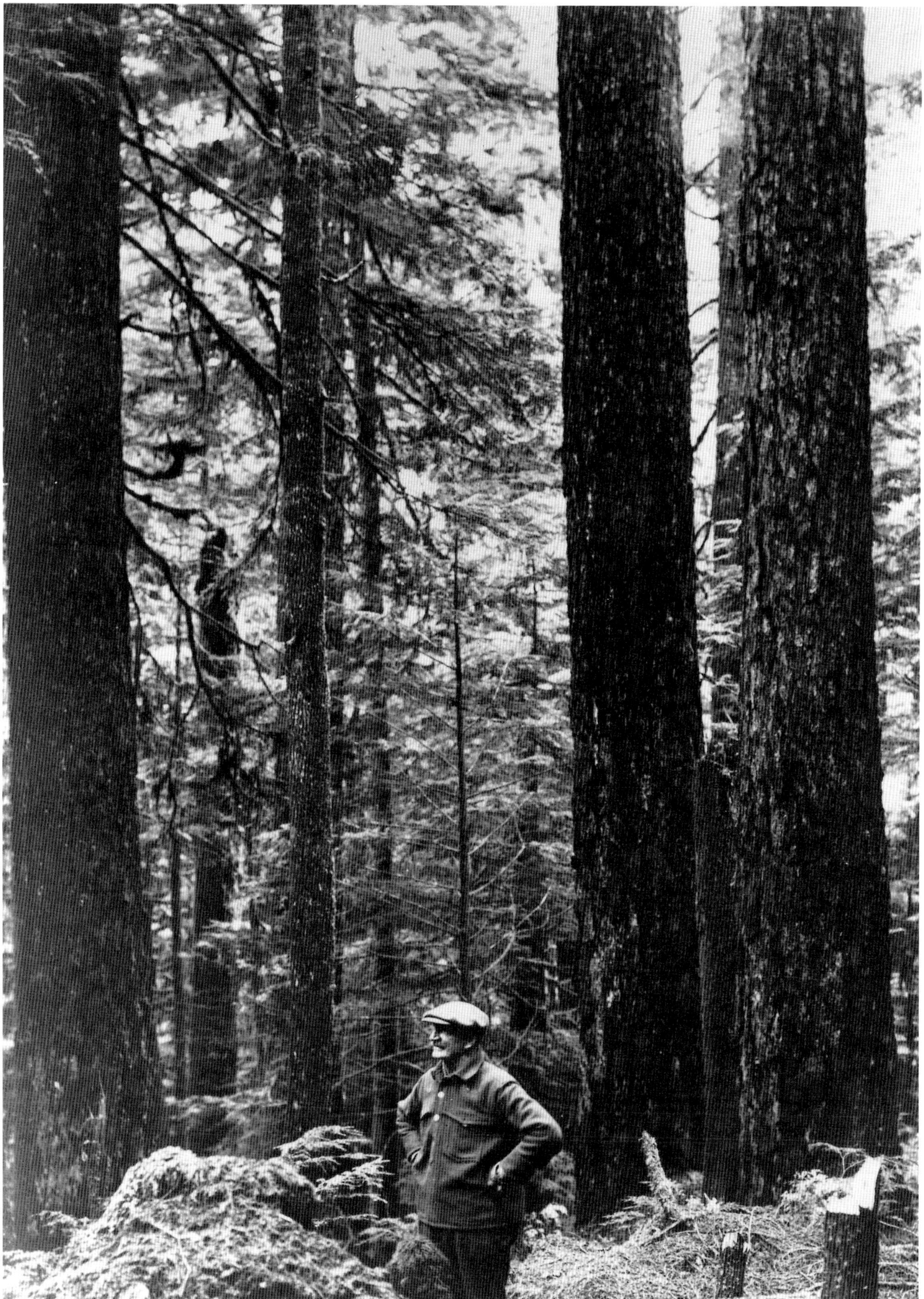

P.B. Anderson's Knox Bay woods camp was one of the biggest independently owned operations on the coast in 1917. (UBC P.B. Anderson Collection)

P.B. Anderson's Knox Bay beach camp with log dump at right. (UBC P.B. Anderson Collection)

P.B. Anderson logging crew, Knox Bay woods camp, 1917. (UBC P.B. Anderson Collection)

One of Union Steamships' coastal steamers at Knox Bay, 1917. (UBC P.B. Anderson Collection)

Later they went up to Salmon River and founded the Salmon River Logging Company.

So, after fifty years of hard slugging, with a never-flagging confidence in the natural resources of the BC coast, P.B. Anderson finally pioneered his way to success. It wasn't easy – there was only one way open to success and that was the hard way. He took it and finally succeeded.

The last time I can remember talking with P.B. Anderson was about 1952, and I can well remember his parting words at that time: "You know Swanson, I'm eighty-one years old but if I could find a nice bit of timber, and get a few old-time loggers and maybe a few horses, I think I'd like to start logging again in a small way, you know. Because there's no fun nohow, in doing nothing."

JOHN BLACKLOCK "DADDY" LAMB

The history of the BC coast is the recorded events of pioneer woodsmen who carved out the first big clearing, starting at the rugged shoreline and pushing it farther and farther back to make room for our ever-growing cities and our agricultural farmlands of today and tomorrow. Such a pioneer woodsman was John Blacklock Lamb, the founder of the Lamb Lumber Company – a man who made two blades of grass grow where none grew before.

It was at Squamish in 1912 that "Daddy" Lamb started letting daylight into the swamps. Some of the logging was ground-lead and some was done with horses. In 1915 when that claim was logged off he moved to Lang Bay. It was in 1920 that the daddy of all home-guard camps was born: Daddy Lamb's camp at Menzies Bay. Many an old-time logger fondly refers to it to this day as *home*, and every one of them remembers old Daddy Lamb and his outfits at Squamish, Lang Bay and Menzies Bay.

Daddy Lamb was born in 1873, in the family homestead back in Little Shemogue, New Brunswick. He started west in 1898 with $25 in his pocket, which got him as far as Ashcroft, BC, so he walked the rest of the way into Vancouver. Now Daddy had heard his dad tell of Tom Tingley in the Cariboo, who had horses. As young Daddy Lamb knew a lot about horses he went to Quesnel. He was lucky, for there he took on the job of driving sixty-seven horses cross-country to the Yukon. At Telegraph Creek he sold his saddle horse to an Indian for five dollars and cut cordwood for his fare into the Klondike, where he arrived in February of 1899. He had bad luck at gold mining, so he took a job on the White Pass and Yukon Railway and saved his money. One day in 1912 he got a letter from his brother Tom, telling him he'd just arrived in Vancouver, and to sell everything he had and come down. Together they'd go into the logging business, for this was the golden land of opportunity.

Now maybe it was Daddy's boyhood training in New Brunswick or it might have been his Scots ancestry, but one thing he couldn't tolerate was waste. He tried to log the small stuff but the mills wouldn't take it, so he hit on a new and novel idea: to clear off some of the logged-off land and plant hay. Then he got a few cows, a few pigs, a flock of sheep and some chickens. One short-stake brakeman quit because Daddy had fitted out a flatcar with a hayrack and the

Lamb Lumber Company crew in front of bunkhouses at Menzies Bay, 1926. (VPL 1447)

Lamb Lumber's Menzies Bay camp in 1926. (VPL 1451)

Lamb Lumber's yarding crew with a wood-fired donkey, 1926. (VPL 1450)

brakeman had been asked to fork hay, for they stopped logging when it threatened to rain and the crew hauled in hay with the locie. Another time a bridge was on fire and Daddy could easily have saved it, but instead rescued one of his lambs and let the bridge burn. Everything was made use of—it was even whispered around that he used up the old haywire for something. A job was up on the board one time in town for a chokerman for Lamb's Camp who could also milk cows—and the job was taken right away, too.

There was an old bull cook in camp whose hobby was looking after Daddy Lamb, and he'd not go to bed until he knew his boss was safely in for the night. If Mr. Lamb was very late he'd bawl him out and tell him it was time he was in bed—might catch a cold stayin' up late. Well, one night Daddy came in quite late with a few big millmen from town and old Jim lit into him as usual, then brought them in a swell chicken dinner. The next day one of the tycoons told Tom Lamb that they had a dangerous man in camp who should be watched, for he had heard him threaten the boss last night using strong language. Tom just grinned and asked: "Did you notice if his

Lamb Lumber donkey and crew. (VPL 1445)

private speeder was warmed up and ready at the boss's door this morning? And were the fires lit and everything ready for him? Well, that's old Jim, he just worries about Daddy Lamb and thinks the world of him and looks after him like a child."

The old Lamb Lumber Company's camp at Menzies Bay ceased operation after Daddy Lamb, the beloved and revered old head of the big home-guard camp, died in Vancouver in 1932. His immediate family was sitting in the hospital and had just received the stunning news, when by way of consolation his niece remarked, "We may feel bad; but old Jim, up at camp, will feel a lot worse than anyone." And it was six months before the faithful old bull cook could bring himself to realize that Daddy Lamb had passed on into eternity.

REVEREND
GEORGE PRINGLE

Reverend George C.F. Pringle felt pretty blue the first night he pounded through the Yuculta Rapids and headed north in an ancient gas boat. He was to take over the Loggers' Mission on the BC coast for the Presbyterian Church. Before this he had been in the Klondike, where he had endeavoured to save the souls of such men as Dangerous Dan McGrew and Sam McGee from Tennessee. He was now ordained to preach the gospel to such high-ball loggers as Rough House Pete and Spike Dillon, men who would far rather go out on a bender than hear the word of the Lord administered in a bunkhouse.

The Reverend Mr. Pringle was blue, too, because he had just returned from overseas after being a Padre in World War One, and he longed for the comradeship of the men he'd known and understood. This new and unheard-of type of man, the logger—how was he to find his way into *their* hearts? But today, after twenty-five years, you boys on the coast who have known Reverend Pringle will recall his name with that deep respect that can never be demanded, but must be earned by a man.

Back in 1920 he was stationed on Texada Island, and travelled to the various camps in a gas boat known as the *Sky Pilot*. At first it wasn't an easy task to land at a camp and say to some hard-boiled superintendent, "I'm a preacher, and I'd like to hold service in your camp tonight." He often got the reply, "This ain't no Sunday school, Mister, but if you want to come, I guess we can put up with you."

The first time Reverend Pringle went into the big camp at Grassy Bay, he dropped anchor out near the boom and waited until morning. When daylight finally broke in the swamp, he went ashore in the dinghy, but he wasn't getting much of a welcome until the bull cook, a fellow by the name of Pea Soup Brodeur, came up and whispered something in the boss's ear. The boss immediately held out his hand and said, "Howdy, Mr. Pringle, had your breakfast yet?" It turned out that the good minister had saved the life of the bull cook up in the Klondike's frozen wastes. That night the boys in the camp listened to Pringle tell stories of the Klondike. Without knowing it, they had heard the gospel preached to them by the logger-minister and they liked and respected him as well.

Next morning the boss took the *Sky Pilot* out in the woods and

Douglas fir at Rock Bay, 1919. (Leonard Frank photo, VPL 3930)

showed the Reverend a bit of high-ball logging. Pringle asked why had that clump of fine big fir trees never been felled. The boss said, "I'll show you why," and, stepping carefully among the towering monarchs of the forest, he pointed up. There in the tree was a hummingbird's nest with unhatched eggs in it. He explained that when the eggs hatched there'd be plenty of time to fall the trees. This, Reverend Pringle told me later, gave him his first insight into a logger's heart.

The dining room at Bloedel, Stewart & Welch's Myrtle Point camp in 1919. (VPL 20738)

The night of the big storm at Lang Bay, back in January of 1921, the Reverend Pringle and six loggers were taking a shortcut through the tall timbers to a meeting, when the hurricane broke loose. Trees fell all around them, and two of the men were injured. When they finally arrived at their destination, they decided to have the meeting in spite of the howling storm outside. The good minister was preaching on Joshua and Jericho's collapsible walls. Although Mr. Pringle emulated the saints in the Good Book, he never expected his preaching to equal that of the wall-shattering days of Joshua. For just then a mighty gust of wind blew the paper off the wall and ceiling, completely covering his small congregation. The windows fell in and the lights were blown out. Well sir, his congregation was convinced right then and there that Reverend Pringle practised what he preached.

The Reverend Pringle, the well-known loggers' minister, left the Loggers' Mission in 1928 to take a church in Victoria, and later he retired on his army pension. Occasionally he officiated at the weddings of his old logger friends or baptized the grandchildren of the couples he married on the *Sky Pilot* in the old days. He was sitting in his wheelchair in his Vancouver home when he told me that the loggers he had learned to love on the BC coast were, in his opinion, the finest bunch of men he has ever been privileged to know. He asked me to give his kindest regards to all of you, and especially his old friends.

Visitors inspect Capilano Timber's Camp A in 1918. (UBC Capilano Timber Co. Collection, Album 33)

The kitchen crew at Brooks, Scanlon & O'Brien's Stillwater camp, 1926. (VPL 1592)

Bernard Timber Company's floating camp at Port Neville in 1927. (VPL 1675)

Floating bunkhouse tent at Bernard Timber camp in Port Neville, 1927. (VPL 1688)

THE HORSEPLAY MURDER

One thing that has always drawn me close to the interesting men I have known during my fifty years in and around the BC woods is their unquestionable sense of humour. Some of them never show the humorous side of life until they are well oiled up; others find humour in their everyday work, while still others never cease to be little boys and their day is filled with pranks and practical jokes played on their comrades. But this fun usually ends in disaster or unfriendly relations. I remember one time in a camp up the coast where the old booby trap was set for me, but the Super walked in and got the contents of a five-gallon bucket of cold water down the back of his neck, which injured only his dignity. I've known bunkhouse cranks to quit or ask for another bunkhouse just to avoid the bunkhouse clowns, but the most drastic climax to practical horseplay I have ever known resulted in a murder at a camp over on the Island. Here's how it happened.

Back in the summer of 1926 they were logging with a Lidgerwood skidder at Camp 9, back of Ladysmith. The chaser was a man by the name of Orville McMann, who had been shell-shocked and was a quiet chap who minded his own business and was fond of stretching out on his bunk of an evening to read his favourite book. Now, the leverman and the second loader—a couple of bunkhouse clowns—shared the same bunkhouse, and proceeded to make McMann's life as miserable as possible. They'd tear pages out of his books, loosen the bolts of his bunk so it would collapse when he'd lie on it, put thistles under his sheets and finally filled his boots with soft yellow cup-grease. He took all these boyish pranks in good part until they carried on their tricks out on the job, where the leverman would open the throttle wide (with the frictions off) when McMann was on the pile unhooking a turn, and then they'd both roar with laughter as he scrambled for his life. At last, one day, after an unmentionably dirty trick, McMann blew up and quit, and as he left the job he warned the pair of pranksters he'd be back some day to get even with both of them.

Nothing was heard of Orville McMann for a long time and his name was forgotten until Slim Trailing, the rigger, met him in Nanaimo on the opening day of hunting season. Orville had his pack and a 38-55 rifle when Slim asked him if he was going out to bag one–to which the old chaser replied, "Yes, maybe two!" The rigger offered him a ride as far as Ladysmith and McMann accepted. No one thought much of it when he rode out on the locie with old Charlie Cathey and slept on the donkey all night to wait for daylight in the big clearing. When the crummy arrived in the morning with the crew, McMann drew a bead on the leverman and said: "You made me suffer plenty, now you're going to suffer." Orville squeezed the trigger. The leverman fell and died on the spot. While this was going on, the second loader took to the woods. McMann saw him disappear so he pumped another shell into the breach and took off after him, and after a while the flabbergasted crew heard two shots. The BC Police searched the timber but neither the murderer nor the second loader could be found.

A Lidgerwood skidder, with loading drums in front and yarding engine at the rear, working at Victoria Lumber Company's Copper Canyon show in 1947. (Jack Cash photo, UBC BC 1930/547/1467)

It snowed heavily that winter, and in the spring the rigger sent the whistle punk out on an old road to locate a receding block and strap. The punk returned white as a ghost and panted, "A man out there—a dead man!" They all went out to see. There was McMann sitting against the tail-holt with the 38-55 between his knees, the muzzle under his chin—and an empty shell in the breach. He'd been there all winter, dead.

The police investigation revealed Orville McMann had no kith nor kin, had cashed in his insurance and blown the works in town with the exception of his fare back to camp. And when his body was searched there was only 40 cents—the exact fare from Nanaimo to the Camp 9 Crossing. Oh yes, the second loader—the next day he turned up, for he knew now he was safe. Thus ended the horseplay for all time in Old Camp 9.

LEN CARY AND HIS FAMOUS SIX-SPOT

The ambitious young fireman stuffed his greasy old overalls into the firebox of the ancient wood burner, Number 21. For fifteen years he'd been ramming slabwood and shingle hay into that infernal hole. But today was his last shift. Tomorrow he would be engineer of the spanking new, 50-ton Shay that had just arrived. His pride knew no bounds to have such an honour bestowed upon him.

The year was 1906 and the place was Chemainus. The ambitious young fireman was none other than the famous Len Cary, destined to become the most photographed locie engineer in British Columbia.

Back in the days when Len Cary started jerking a throttle, railroading was a far cry from the conditions that exist today on our modern logging railways. The logging industry used disconnected trucks with hand brakes. The engineer would whistle for brakes and

Victoria Lumber & Manufacturing Company's ground lead donkey hauls in a 120-foot fir log, with a 26-inch top diameter, 1907. Engineer Len Cary is in left foreground. (BCARS 53600)

Victoria Lumber & Manufacturing Company's Shay at the Chemainus mill in 1937. (Wilmer Gold photo; IWA Local 1-80)

Victoria Lumber & Manufacturing Company's No. 9 Shay taking on water at Chemainus in 1937. (Wilmer Gold photo, IWA Local 1-80)

the brakeman had to climb over the loads and apply the brakes on the fly, by hand. The locie had a steam jamb. Somehow they managed to bring long trains down ten-percent grades, and Len Cary could do just that with the best of them.

One time back in 1908 Cary was easing his six-spot with a train of logs down a long hill. At the bottom of this hill there was a run of adverse grade and a curve. The train got out of control and the crew jumped. Cary was the last to take to the weeds but before doing so reversed the engine and opened the throttle. Away went the train roaring down the hill with only a phantom crew aboard.

After the live crew, including Cary, had picked themselves out of the ditch they started running down the hill to look for the runaway train. They met it snorting back up the hill toward them. It had come to a stop with the reversed throttle wide open and with lots of steam. So it had started back on its own. Cary climbed into the

Victoria Lumber & Manufacturing acquired this No. 9 Shay in 1929. (Jack Cash photo, UBC BC 1930/549/94VL 4020)

This much-rebuilt Porter locomotive was built for Timberland Development Company in 1924 and was one of the last steam engines to operate in BC, for MacMillan Bloedel. It replaced Len Cary's Six-Spot. (Jack Cash photo, UBC BC 1930/549/94VL 4019)

cab and dryly remarked that it was just a matter of having his engine well trained.

When inspection and certification were instituted on the logging railroads in 1920, Len was up on the carpet to be examined on air brakes. The inspector asked Len to explain how an air brake functioned. Cary replied, "It's this way. You see, zip, she's on and zip, she's off. That's all there is to it." Then the inspector threw the book at the engineer. He got the works; but he also got his ticket because he could handle a locie.

The company decided after about twenty years that Len's old six-spot was too small for the heavy grades in the woods. Len's heart was nearly broken. He went to the boss and talked like a Dutch uncle until it was agreed that old six-spot could switch the mill yard and Len could run her. He'd been running that locie so long he nearly owned her. And as the years wore on, Old Len Cary and his old wood-burning locie with the balloon stack became almost legendary figures around Chemainus.

Len would polish her up and spot her near the Island Highway, and tourists would stop to have a look at the comical-looking locie. Old Len would tell them she was at least a hundred years old; that she'd come 'round the Horn in a windjammer and was the first locomotive ever to run in BC. They would swallow his story hook, line and sinker, and would get him to pose for a picture standing alongside the relic. Len became famous.

Some of the most sociable beers of my life were had in Len Cary's company, sitting in the Horseshoe Bay Inn at Chemainus. I'd often drop in to have a chat with the old-timer on my way up the Island. He had a cabin nearby where he'd lived alone for many years. Well, not quite alone – his beloved six-spot was parked alongside the cabin door, gently sizzling as old locies are wont to do. Many's the time he's taken me out and proudly displayed the old girl to me by the light of a brakeman's lantern. Old Len was so attached to the ancient locomotive that a legend arose to the effect that when Old Len passed on his locie would be retired from service.

One night I dropped in as was my custom to see Old Len. He wasn't around. Next day I saw an oil-burning locie running around the yards. It remotely resembled Len's old six-spot. The boys told me that Len hadn't been feeling well and was taken to the hospital. When he was told he would never be well enough to run his old locie again he seemed to lose interest in life. And so he passed on to the Land of the Heavenly Timber.

As a tribute to his memory the boys discarded the balloon stack and modernized the old six-spot, converting her into an oil burner.

Thus ended the career of Vancouver Island's most famous locie engineer. Born in England in 1880, Len died in Chemainus in 1946, and he was buried within whistling distance of where his old six-spot was still wheeling in logs in Chemainus. Eventually she was replaced by the 1044, and at this writing both Len Cary and the six-spot are now a part of Chemainus history.

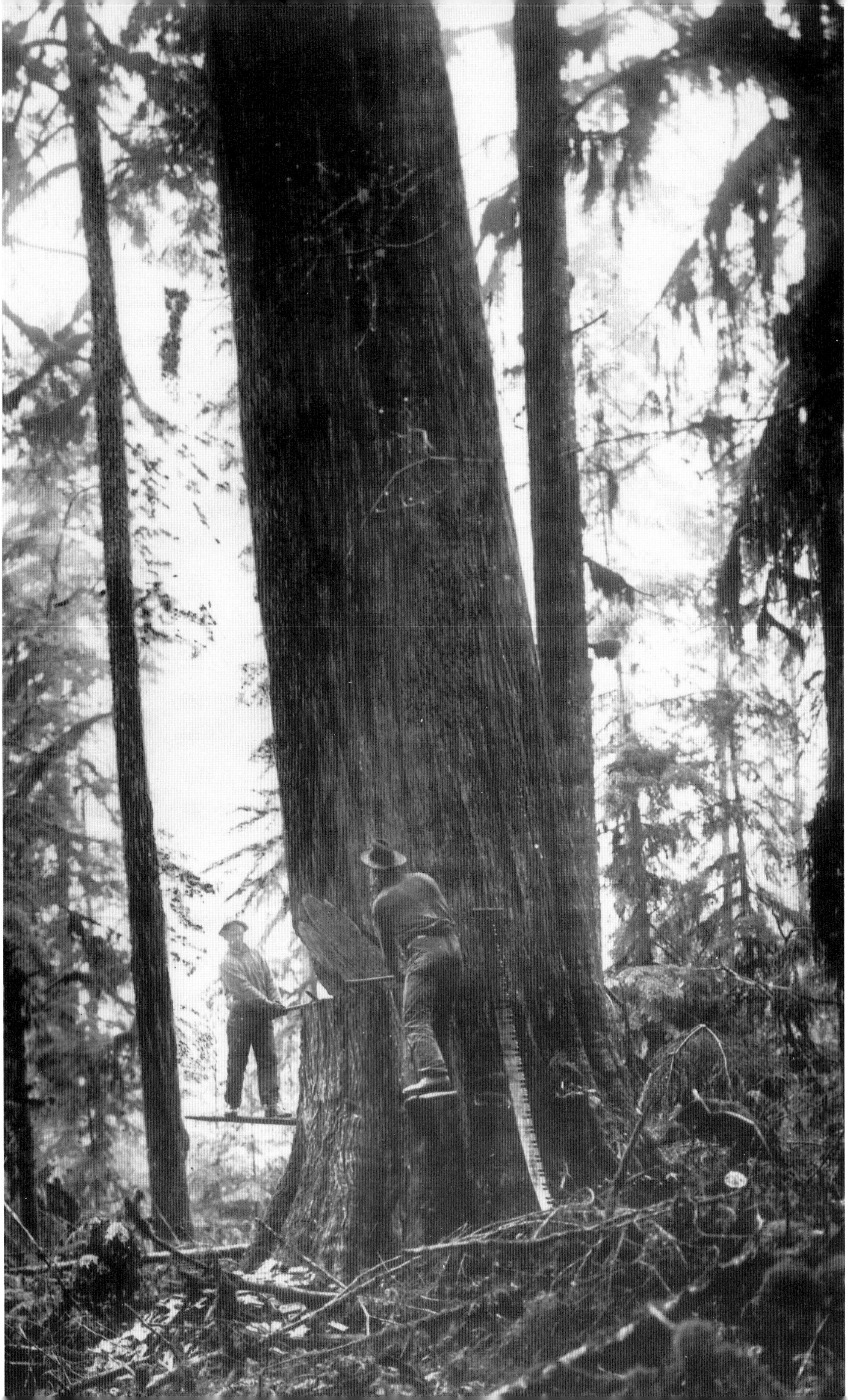

BILL BARBRICK

Bill Barbrick, the bull skinner, led his cattle back to the barns one rainy morning in 1887. They were logging off the present site of the Vancouver General Hospital. One or two of the old-time loggers had come out on the skid road and, after sizing up the weather situation, had declared there would be no work as "it looked like it might rain all day." It was Saturday anyhow, and they would all walk down to Gassy Jack's where there was lots of beer to be had.

The bull skinner felt relieved because the yokes made his oxen's necks sore if they worked too long in the rain. He handed the prod

Opposite: Handfalling in Vancouver in 1890s. (CVA TR P60 N49)

Below: Logging with oxen on Granville Street in Shaughnessy in the 1880s. (VPL 30172)

Opposite: A high-lead yarder with duplex loader at Bloedel, Stewart & Welch's Myrtle Point operation in 1919. (VPL 20744)

Brooks, Scanlon & O'Brien's boom crew at Stillwater, 1926. (VPL 1569)

back to the stable boy and joined the loggers who were walking down the skid road. That skid road is today Oak Street. They crossed False Creek and took the trail to Gassy Jack's.

To Bill Barbrick, bull skinning was no new game. He had arrived recently from the Panama Canal Zone where they logged teak and mahogany. Bill had prodded bulls in the tropical heat of Central America. While Douglas fir was new to him, a bull was a bull in any country. Bull skinners the likes of Bill Barbrick were few and far between.

William Henry Barbrick was born in Halifax, Nova Scotia in 1860. He logged with horses around the Bay of Fundy in 1887 and went to Jamaica Plains in 1880. Before coming to BC in 1887 he had knee-bolted in a shingle mill in the state of Washington. He actually worked sixty years in the BC woods.

Old Bill Barbrick was a character if I ever saw one. The last time I talked to him was up at Menzies Bay in 1947. He asked me into his little shack (by this time Bill was too much of a character to live

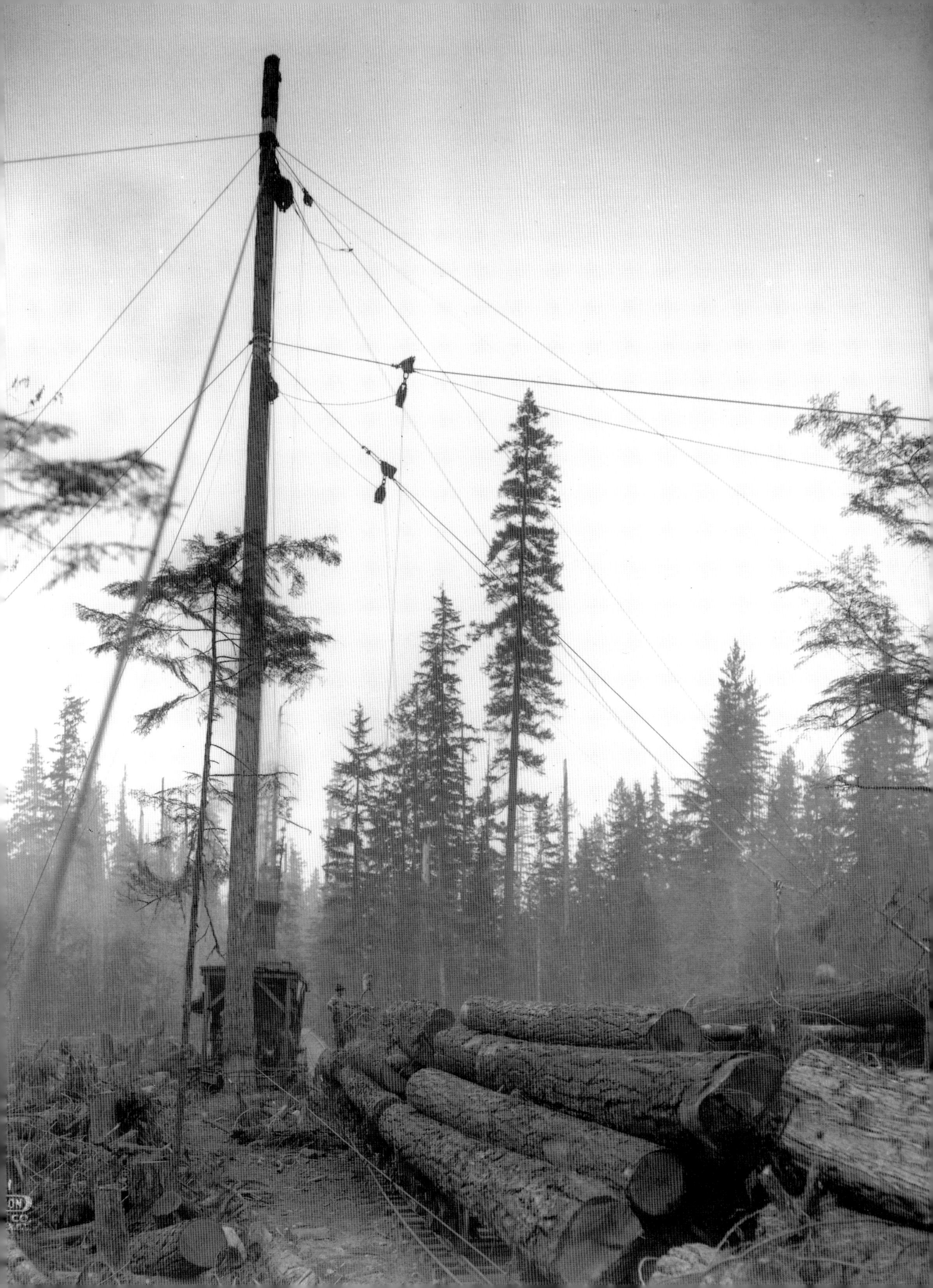

Duplex loading onto skeleton cars at Bloedel, Stewart & Welch's Myrtle Point operation, 1926. (VPL 1513)

in a bunkhouse), shooed a litter of kittens out of the way and asked me to make myself at home while he recounted the story of his life.

Bill told me that when the bull teams all died off he took to working on the boom. From there he got to be mate on a barge, the *Sea Wolf*, bringing logs from the Queen Charlotte Islands. That was back in 1912.

One time Bill was shipwrecked when a storm blew up on Hecate Strait. The *Sea Wolf* was in tow loaded to the gunwales with logs. The tug strained at her towline with the *Sea Wolf* dragging both anchors. They were losing ground when finally the tug whistled to cut loose. The *Sea Wolf* was cut adrift with Bill Barbrick and ten other men aboard. She piled up on the beach at Dead Tree Point but Bill and the crew took to the lifeboats and got ashore. The night was as black as the inside of a cow as the survivors wandered, soaked to the skin, through the dense underbrush of the Charlottes. Somehow they stumbled onto a trapper's cabin where they got food and shelter.

It was believed they all perished. But after several days Bill located the wireless station and sent a message they were all okay. He was instructed to put a watchman on board the *Sea Wolf*. But when he got there it was too late. The Indians had taken everything, including the hinges on the captain's door. The barge was declared a derelict. A boat was sent to pick up the survivors and Bill went back on the boom at Myrtle Point for Mr. Riley.

"Yep," said Old Bill, "it was in 1911 that I first worked for BS&W, at Myrtle Point. That was the place where Bloedel's rigged their first spar tree. Nobody thought it would work at first but I see they're still using 'em. That was the place, too, where Mr. Riley invented the first duplex loader. He got the idea from the log winches on the old *Sea Wolf*." Thus Bill told his tale.

After thirty years of faithful service with Bloedel's, Bill gradually became a home-guard. When he reached the age of seventy-nine the company retired him on a pension. But he didn't like the idea. He wanted to feel he was still a cog in the wheel of the great

Duplex loading at Victoria Lumber & Manufacturing in 1940. One of Bob Swanson's first jobs in the woods was running a duplex loader. (Jack Cash Photo, UBC BC 1930/549/92VL 4016)

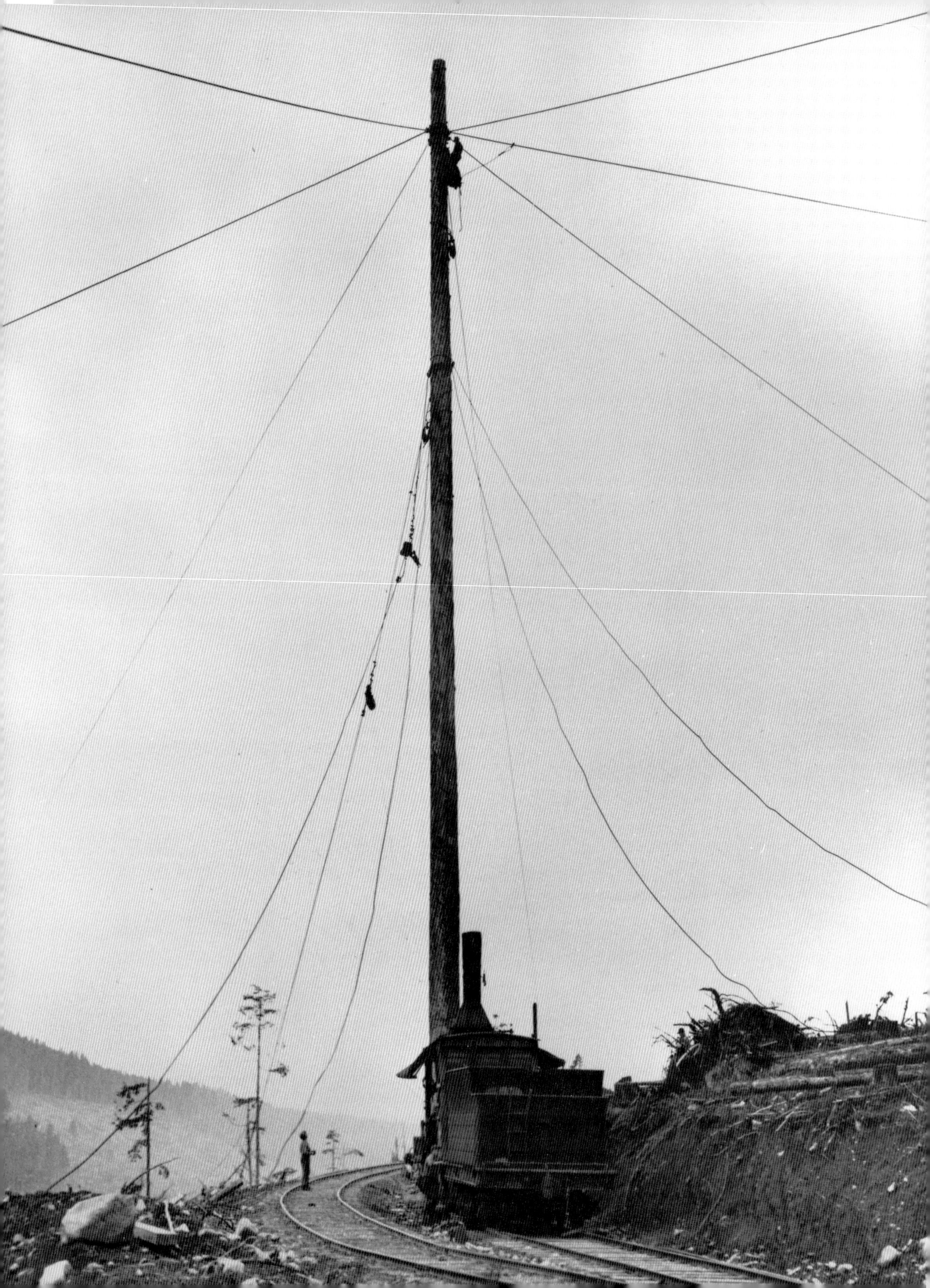

company he had helped to build. He longed for the thunderous rumble of logs as they tumbled off the cars into the salt chuck. He also had a nostalgic feeling for the rhythmic beat and musical chime of the cookhouse dinner gong. He missed Karlson, the cook at Menzies Bay, and Jack Smith, the locomotive foreman. The solitude of the city was getting him down.

Finally Old Bill could stand it no longer. He took the matter up with Sid Smith who instructed camp to fix up a little cabin for Old Bill. He sent Bill back to Menzies Bay to do just as he liked – on company pension. There was one stipulation that Sid made: that if anything happened to Old Bill he was to be immediately notified.

Old Bill amused himself with a little garden and a few litters of kittens. He also took on the duties of looking after the company's oil tanks. On last seeing him in 1947 I asked him if he was happy and really his own boss, to which he replied, gently stroking his long white beard, "Well, yes, I guess I am, outside of one thing. That is that these young superintendents yell at me to go home and put on my hat as I might catch my death of cold. Outside of that I guess I'm my own boss."

The last time I was up at Menzies Bay, in 1956, I learned that Old Bill Barbrick passed on to the Camps of the Holy Ghost a couple of years before: he was eighty-seven years old and active and happy to the very end.

Opposite: Rigging a spar tree for a skyline show at Victoria Lumber & Manufacturing in 1940. (Jack Cash photo, UBC BC 1930/549/65VL 4003)

"BROOMHANDLE" CHARLIE SNOWDEN

Any time you meet a man who can tell you from memory what happened on Vancouver Island back in the early eighties, well...you can figure you are talking to a pioneer. If that old-timer happens to be a friend of yours, you'll hear real tales of the early days. A few years back when I happened to be in Nanaimo, I ran across just such an old man. His name was Charlie Snowden. I hadn't seen old Charlie for about twenty years, so we sat back and nursed a few hot rums while we went back over the years together.

Charlie Snowden was born in England in 1874 and arrived in Nanaimo on August 25, 1876. That was before the CPR was built. With Charlie a babe-in-arms, his parents made part of the journey across America in a covered wagon. After arriving in San Francisco they embarked on a sailing ship for British Columbia.

If Charlie had attended school like a good boy he would still have turned out to be a pioneer—but not the colourful character he is today at the age of seventy-four. For back in the early eighties he would play hooky and steal rides on the narrow-gauge railroads that ran from Departure Bay out to the Wellington coal mines, where his dad was engineer. At the age of fifteen Charlie left school to go braking on the little coal trains. When he was seventeen he was running the locie himself.

The first locomotive he ever ran was called the *Pioneer*. The *Pioneer* was also the first locomotive in the colony of Vancouver Island. She was built in England in the late fifties and weighed ten tons in working order. Other little locies around Nanaimo were called the *Uclataw*, *Nanaimo*, *San Francisco*, *The Duke*, *The Duchess* and the *Robert Dunsmuir*. Charlie Snowden was a proud boy as he wheeled those little brass-trimmed engines over the railroad that is today Nichol Street in Nanaimo; or as he pulled a train of lumber over the narrow-gauge railroad that ran from the Mill Stream Sawmill out at East Wellington, down to the wharves at Departure Bay, back in 1892.

Charlie can remember, too, when the Esquimalt & Nanaimo Railway made its first trip into Nanaimo in 1889. Another red-letter day in his life was the day in 1897 when he switched 6400 tons of coal in six hours and thirty-five minutes, thus beating the coal-loading world record in those days.

It was in 1906 that Charlie Snowden started to punch donkey in the woods over on the Island. They were loading with horses, and the yarder Charlie was punching was an old "BC" 10x12, built by the Albion Iron Works. At first the push didn't think much of the new donkey puncher, as he was too careful. It wasn't until Charlie had broken the mainline and all the chokers that the boss swore he was the best donkey puncher ever to jerk a throttle.

John Coburn built his first sawmill in 1907 at Cassidy and bought Charlie's old locie, the *Nanaimo*, to do the hauling. It had been running after a fashion for a few days, but wouldn't pull a hat out of a lard bucket. They sent for Snowden. After telling Coburn he'd cut his eye teeth on that throttle, Charlie jumped aboard, tooted the whistle, and wheeled the train of lumber out to Coburn's siding in grand style.

He made such a hit with Coburn that when the new Ladysmith Lumber Company was built at East Wellington in 1908, Charlie was picked to run the locie and the railroad. This locie, too, was of historical interest as it was number 178 from the New York Elevated

Royal City Planing Mills used this mainline locomotive to haul logs from its Surrey operation along the New Westminster Southern Railway line to Port Kells in 1899. Robert Harvie was the engineer. (CVA Log P55 N39)

A ground-lead steam yarder with a horse on the haulback line at Victoria Lumber & Manufacturing Company's Camp 2 near Ladysmith in 1902. (BCARS 53608)

Railway, and was built in 1865. Even though it had a weird type of vacuum brake, using a bellows, Charlie managed to bring long trains down heavy grades with the 17-ton engine.

The 178 finally became so haywire that the pioneer hogger had to take her to the E&N roundhouse at Wellington for a complete overhaul. When the repairs were completed, there was a leak in her water tank and the E&N boss had to drain her. He looked inside the tank and saw it was stuffed with most of the shop tools. The "brass hat" was pretty sore at Charlie for trying to make off with his tools, and Charlie was brought up on the carpet to answer for his sins. He told them he would have put the wheel lathe in there too, but he couldn't lift it. So they let the matter drop at that.

In 1914 the 178 coughed her last and was replaced with a brand-new 27-ton saddle-tank job which us kids nicknamed "The Coffee Pot." At the same time we dubbed Charlie "Broomhandle Charlie." When we used to steal rides, Charlie would sock us with an old greasy broom which he kept soaked in a bucket of oil for our benefit. Many's the walloping I've had for my own good from old Broomhandle Charlie. He kept a shotgun on the engine, too, so he could bang the odd pheasant along the road. One old farmer got wise to him and threatened to prosecute him for climbing over his fence to pick up the birds. Charlie pleaded he didn't climb over the fence (he climbed under it), and the next day a few tons of coal "accidentally" spilled from a coal car right alongside the farmer's woodshed. Charlie and the farmer were friends from then on.

It was in the summer of 1919 that I first worked with Charlie Snowden, although I'd known him since I was three years old. I still had a fear of his terrible broomstick as I rode on the pilot of his engine and hand-sanded the rails on ten percent grades. We were logging on Mount Benson, back of Nanaimo, and old Charlie wheeled in the logs, switched the mill and hauled the coal from the famous Jingle Pot Mine.

When the claim was finished in 1930, the old hogger took a job as engineer at the Nanaimo Hospital, a job he held until he retired at

Victoria Lumber & Manufacturing Company's No. 3 Climax at Camp 5, west of Ladysmith, 1907. (BCARS 53605)

Parbuckle loading at BC Mills Timber & Trading's Rock Bay camp in 1913. (Leonard Frank photo, VPL 5796)

the age of seventy-two. This ended an active career of fifty-eight years. When I parted company with the old pioneer and his wife a few weeks ago in Nanaimo (after the rum was all gone), he told me the happiest days of his life were the days when he used to chase us kids off the pilot of his locie with that greasy old broom handle.

HARRY SMITH
THE TIMBER CRUISER

Harry Smith, the timber cruiser, wondered how long it was going to take him to understand the Haida tribes of the Queen Charlotte Island Indians, for he'd offended them again by pitching his tent on the sacred burial ground of the Wolf Tribe of Haidas, and Captain Billy Brown, the Chief of the Tribe, had arrived by canoe and protested to the timber cruiser that the sacred spirits of his Wolf ancestors had been offended. Harry appeased him with gifts of tobacco, re-pitched his tent and sat down to talk it over with the Haida Chief who asked Harry, "What for white man come?"

"To count big trees for the Great White King," was the timber cruiser's reply.

"Me help too," said the Chief. "Tomorrow show you many fine canoe trees." And so Harry made peace with the Chief and hired a crew of Haida helpers whom he agreed to pay in Hudson's Bay blankets.

This happened back in 1911 when Harry Smith was sent to cruise the virgin timber of the Queen Charlotte Islands, and he'd crossed Hecate Straits with his party in a forty-foot Haida canoe. Harry was then twenty-five years old and had cruised timber in all parts of BC since he was a boy, but the Sitka spruce and the Haida Indians of the Queen Charlottes were a new and fascinating experience to him.

It was in 1943 that I first became associated with Harry when we worked together on spruce production in the Charlottes, where he was, after thirty-five years, still cruising the big spruce. By now he had learned to know and understand the Haidas, for he was not only an expert woodsman but a pioneer diplomat as well. Many's the hair-raising tale of old times that was spun by Harry as we sat and smoked around the bunkhouse stove.

In the early days Harry wandered into many deserted Haida villages. His crew of Haida braves hung back, explaining that the deserted villages, with their towering and decaying totem poles, were taboo. A great plague had swept the Charlottes years before, and the towering forest had again defeated the Natives.

Harry took me back into the timber one day and showed me the old skid roads over which the Haidas had dragged their newly made

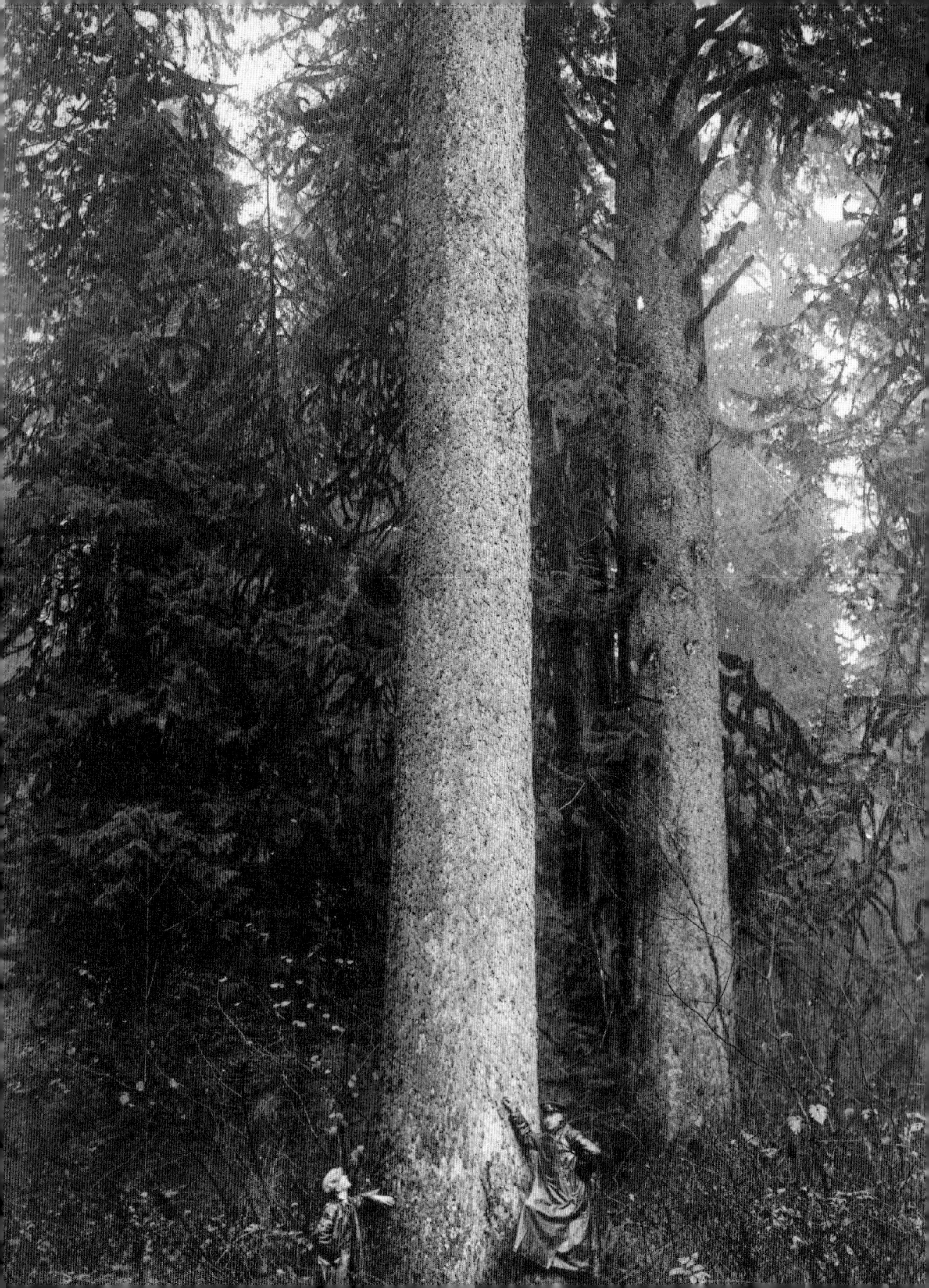

Timber cruising camp on the Queen Charlotte Islands, 1954. (UBC BC 1930/301)

canoes two miles to the salt chuck. He showed me several partly completed canoes which had been abandoned and had never been taken out. One of them was forty-four feet long and grown over with moss and small hemlock trees. We chopped down one of the trees growing in the canoe and counted the yearly growth rings. It had been growing in the abandoned canoe for 135 years! Seeing that I showed such an interest in all this, Harry took me over one Sunday to meet the eighty-year-old Haida Chief, Captain Billy Brown, who gave me firsthand information on the early days of Haida logging and canoe building.

The Chief said that 200 years ago his people had fashioned axes from the iron spikes of stranded windjammers, and before this they had used stone axes and felled the cedars by burning them down. The first step in building a war canoe was to chop a hole into the heart of a big cedar and, if it proved sound, it was felled; if not, it

Opposite: Timber cruising in a Queen Charlotte Islands spruce stand in 1918. (VPL 3869)

Winter cruising in the Queen Charlotte Islands, 1918. (Leonard Frank photo, VPL 3863)

was left standing. Many of these test holes are to be seen in the cedars of the Charlottes to this day.

After the tree was felled and chopped off to the length of a canoe, and if it still proved sound, the canoe maker then brought his family and set up camp, after which they fashioned and hollowed out the canoe. This would take nearly all summer.

In the fall the canoe was covered with a roof made of cedar shakes, and it was left in the woods one winter to season. Should a defect be revealed or should the builder die, the canoe would be abandoned and never taken out. In the spring, if everything had been favourable, the Chief and his tribe would journey out the skid road to the canoe and hold a big feast in its honour, after which, with the aid of ropes made from spruce roots, the tribe would chant as they dragged the new canoe to tidewater. It was then towed to the home village, where it was fashioned by master craftsmen until the sides were one and a half inches in thickness, then placed on two

Timber cruisers built this split cedar food cache on the Queen Charlotte Islands. (UBC BC 1930/301)

logs and filled with hot stones, covered up, and water poured over the stones. The steam made it possible to spread the sides and give the keel the necessary bow. The warhead, paint and scalp boards were put on and the canoe was finished.

Billy Brown then told me of the big canoe that couldn't be taken out. It seems that this one was twelve fathoms in length and was too large for the one tribe to haul on the skid road, so another tribe was asked to help. After the customary feast and a big pow-wow, the Chief could see the other tribe was jealous of the big canoe and, as this was considered an evil omen, he decreed it should be left in the woods and called "the canoe that couldn't be taken out." I could see that Harry Smith grinned at this, and afterwards he told me it was no fable – he had located the canoe on the Ain River where it is still covered with a cedar-shake roof.

The last time I saw the old Haida Chief he invited me to his home, where he sang me a Haida war song and presented me with a stone totem pole and a shell ashtray, and he remarked on how it grieved him to see the old Haida customs being replaced by the white man's way of living.

The last time I saw my old friend Harry Smith the Timber Cruiser was the other day on Granville Street. He was as hale and hearty as ever and told me he'd been cruising timber by aeroplane. His farewell salutation was: "If I locate any more Haida canoes, I'll let you know."

THE HANDLOGGER'S CHRISTMAS

The old handlogger buttoned up his mackinaw and breasted a northeast wind as he trudged over the trail that led to his little log cabin at the head of the bay. The warmth of his old tin stove would feel pretty good tonight. It seemed to him, though, the snow had fallen early for this time of the year. But then, after all, time didn't mean much in a place like Loughborough Inlet. It was easy to lose track of time, and it didn't matter much anyhow.

He felt good to think that he'd got the last of his logs into the water before the snow was too deep, yet something was troubling him and he didn't feel right. Maybe it was the argument he'd had with old Hank last summer about those beachcombed logs. Hank wasn't such a bad old fellow, he thought. It used to be nice to go over to his cabin once in a while and have a game of crib, but since the big argument they weren't even speaking. Oh, it was Hank's fault, all right, but then, the old handlogger thought to himself, maybe he was just too cussed himself to admit he was wrong. Yes, maybe he was. Strange why he should feel like forgiving that old neighbour of his tonight. Oh, he'd feel more like himself after he'd had something to eat.

As he rounded the point he could hear the breakers pounding on the rugged shoreline. A four-prong buck, standing on the knoll, was startled by his approach and bounced across the trail, only to disappear under the snow-laden branches of a hemlock thicket. Normally the old handlogger would have gone for his 30-30 and bagged that beauty, but even that seemed wrong to him tonight.

Well, there was his cabin, his little log home, dwarfed by the two towering cedars grotesquely guarding it with their drooping evergreen branches now white with the winter's first heavy snowfall. He'd never seen it look so picturesque as it did tonight, just like those fairy tale pictures when he was a boy.

Even after he'd had his supper and lit his pipe he didn't feel right. It was strange how his mind lingered tonight on his bygone years. Even the scenes of his boyhood seemed but yesterday as they flashed across his aging memory. How well he remembered those happy days at the old homestead, back in Nova Scotia – the split-rail fences poking their weatherbeaten tops from under a blanket of snow... the naked oaks waving their gnarled and twisted arms in the scowling Atlantic wind. He used to imagine they looked like forlorn vagabonds in the winter twilight. And the old grey team, Gerry and Dexter, with their flanks steaming and the sleigh bells jingling their crisp and merry tune as they plodded along the country road that led to the old homestead. Strange how he should remember, tonight, those sleigh bells of long, long ago. And his brothers and sisters, all ten of them. He wondered where his younger sister was tonight—it had been twenty years since he had heard from her.

Now, what was that fellow's name that drove team back in Michigan? McDonald? That was it—Angus McDonald. They had come out together on the spring drive in '86 and had blown their stakes in Duluth—it was a hard winter, '86—and he and Angus had beat it West on a freight. Never heard of Angus again after he went to the Klondike.

And Idaho—that was a long time ago, too, since he'd sat on the deacon's seat and heard the old bunkhouse tales over and over again. But this row with old Hank last summer—that seemed to keep bobbing up and sticking in his throat. Yes, maybe Hank was right after all. It would be nice if they were friends again at that!

The old handlogger polished the bowl of his pipe against the side of his nose as he pondered it all over. It had stopped snowing and maybe a walk in the crisp night air would do him good. But he wouldn't go over to Hank's. Much as he'd like to, his pride wouldn't let him do that. He consoled himself with that thought as he stepped out into the clear frosty night.

Looking over a handlogging show at Loughborough Inlet in the 1920s. (VPL 3782)

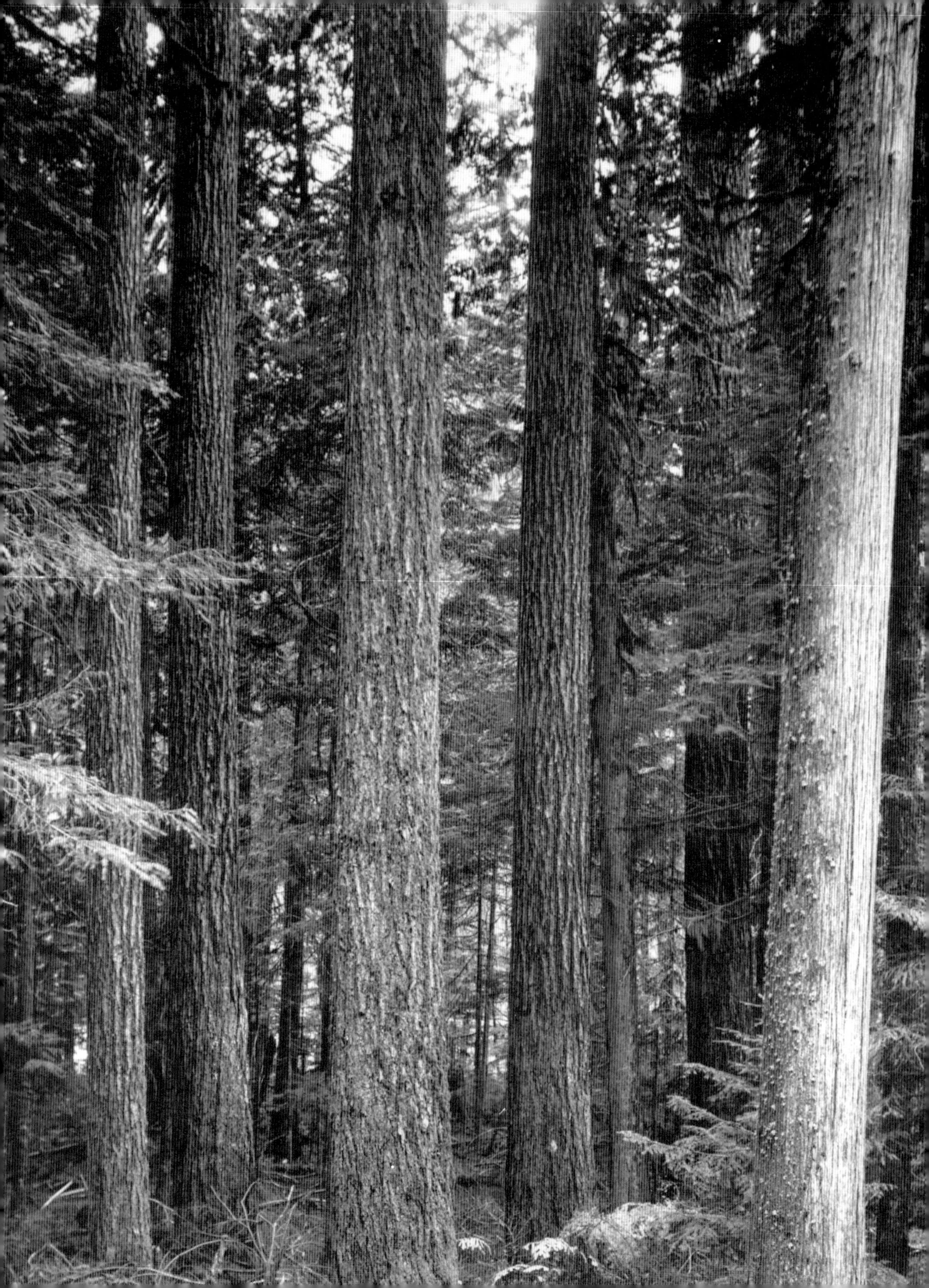

To the old handlogger snow had always been something he'd had to fight to make a living from the potentialities of nature's forest; but tonight his very being seemed to respond to the sheer beauty of the forest land as it sparkled with its myriads of jewels of frozen fire. There, alongside the trail were Douglas firs, like sentinels of the woodlands, pointing their spire-like pinnacles to the zenith of the heavens, while the heavily laden branches of the saplings swayed as they silently slipped their burdens of snow to the ground. The clouds had cleared and the old handlogger could see the constellations of stars scintillating like blazing diamonds on the velvet background of the sky, and as he aimlessly wandered along the trail they seemed to be moving along with him. He stopped, and there in the East was a star whose dazzling brilliance eclipsed the splendour of all other stars in the heavens. That star struck a note on the heartstrings of the old handlogger for, strangely, all of his animosities vanished as he suddenly realized it must be Christmas Eve. That was it, that's why he had been so troubled. But here he was now at the very door of old Hank's cabin and he didn't even remember walking over.

As he stood there wondering how this had happened, the door of the shack swung open and there stood another old man, bowed with years of labour, but sturdy and defiantly strong still. "I was kind of expecting ye, old-timer," the figure said as he stood aside to let the old handlogger in. "So happens I put some coffee on and got the old crib board out. Merry Christmas to you."

Unconsciously, unknowingly, like a winged spirit, the peace that passeth all understanding had reached out to that distant, deserted inlet where two human beings lived in solitude with only sentinel firs for friends... and as they too joined in the Christian spirit of peace on earth, goodwill amongst men, the spirit passed on. But the star burned as bright, and with its radiant beam it seemed to send a message to the old handlogger and his friend, hunched over their worn crib board and well-thumbed cards: a message of satisfaction, an invocation to God rest you merry gentlemen, let nothing you dismay.

Mixed stand belonging to Capilano Timber Company, 1918. (UBC Capilano Timber Co. Collection, Album 33)

LOCIES LIVED AND BREATHED FOR INSPECTOR JACK SHORT

Jack Short, the newly appointed Boiler Inspector, wondered just what sort of men loggers really were one stormy morning back in 1920. He had been sent from Vancouver into the East Kootenay District to inspect a logging locomotive. The "walking boss" of the camp had loaned him a horse and told him to follow the trail on horseback. He'd find the locie, said the WB, waiting for him in three feet of snow at the end of the fifteen-mile trail.

To Jack Short, locomotives had air pumps and boilers and they lived and they breathed. But a horse! That was a locomotive of a different colour. He got there all right and inspected the locie while Dobbin waited. But after the return trip (by horseback) Jack Short did his wondering about loggers, and he did it standing up.

Locomotives had always been Jack Short's first love. He started his career fifty-five years ago in Stratford, Ontario, where he served his time as a locomotive machinist. From there he went to Chicago and in 1894 went railroading in Mexico. In 1900 he was Locomotive Foreman at Devil's Lake, North Dakota. Being a boomer he moved farther and farther West until he landed in Vancouver on a freight train one cold November morning in 1902.

He was riding the tender of a CPR locomotive, heading east out of Vancouver one night. The engineer, a fellow by the name of "Smoothy" McMillan, said to him, "Young feller, you look cold. Can you shovel coal?" To which Short replied he sure could. After sharing the engineer's lunch, he fired the engine right into Revelstoke. There he slept on a bench in the roundhouse until it came daylight. Times were hard but the foreman put him to work next morning, for good machinists were hard to get even then. Short became so popular with the boys that he stayed on that job for sixteen years and worked his way up to an air brake expert.

It was in 1920 that most of the BC logging camps were discarding hand brakes and installing Westinghouse air brakes. And where could the industry find a man better versed in locomotives and air brakes than Jack Short? So Jack was hired and for many years travelled into every logging camp in BC. He earned the reputation of not only being an expert locomotive man but a good fellow and

Logging crew at bunkhouse of Victoria Lumber & Manufacturing Company's Camp 10 on Cowichan Lake, 1926. (VPL 1432)

a square-shooter as well. I know every logger who has known him will fondly remember him as Honest John Short.

The first time Jack Short visited Camp 10 at Cowichan Lake he spent nearly a day trying to find the superintendent. Several messengers had been sent out to the Super, telling him that Mr. Short was looking for him. But it appeared that the Soup was avoiding Jack. Finally Jack met a man walking around the front of the locomotive and he said to him, "Say, my name's Jack Short and I'm trying to find the superintendent." To which the amazed, well-dressed logger replied, "So that's it? It's not a gag after all! I'm the Super, and my name's Jack Long." And Jack Short and Jack Long shook hands and became good friends right then and there.

Another time up at Campbell River, Mr. Short got lost and spent most of the night wandering around out in the timber. Here's how it happened: Jack had finished his work at the old CRT camp. He had been phoned and told that he was urgently needed over at Bloedel's. It was about thirty miles to Bloedel's by railway but there was a shortcut through the timber that was only about two miles. After supper, Short got his grips. He received detailed instructions

An Industrial Timber Mills Ltd. Climax at Camp 6, 1934. (Wilmer Gold photo, IWA Local 1-80)

One of International Timber's camps near Campbell River, 1926. (Bill Roozeboom photo, VPL 1459)

from Hank Phelan how he should go so many sections north and then bear to the northeast so much, and then he started out in search of the other camp.

About midnight Hank got a phone call that Short hadn't arrived. Maybe he was lost. About two o'clock in the morning, Jack arrived back at the same camp he had started from, thinking that he had found Bloedel's. He explained he was sorry he was so late. He had met so many bears on the trail, he said, that he didn't care if he ever saw another bear again as long as he lived.

Jack Short retired in 1946 for a well earned holiday, and his friends will be pleased to learn he enjoyed life for a good while after that. One day when I called on him, he said to say hello to all of his old logger friends in the woods, especially the trainmen handling heavy trains on eight percent grades, and those who would have to learn to drive a logging truck.

International Timber's Camp 4, 1921, with No. 1 Shay on left and No. 4 Shay on right. (CRM 11133)

A Lake Logging steel gang laying track near Rounds in 1937. (Wilmer Gold photo, IWA Local 1-80)

The crew of a wood-fired cold deck yarder at Hillcrest Lumber's Sahtlam logging division, 1940. (Wilmer Gold photo, IWA Local 1-80)

Opposite: The Victoria Lumber & Manufacturing bridge crew building the Jones Creek trestle in 1937. (Wilmer Gold photo, IWA Local 1-80)

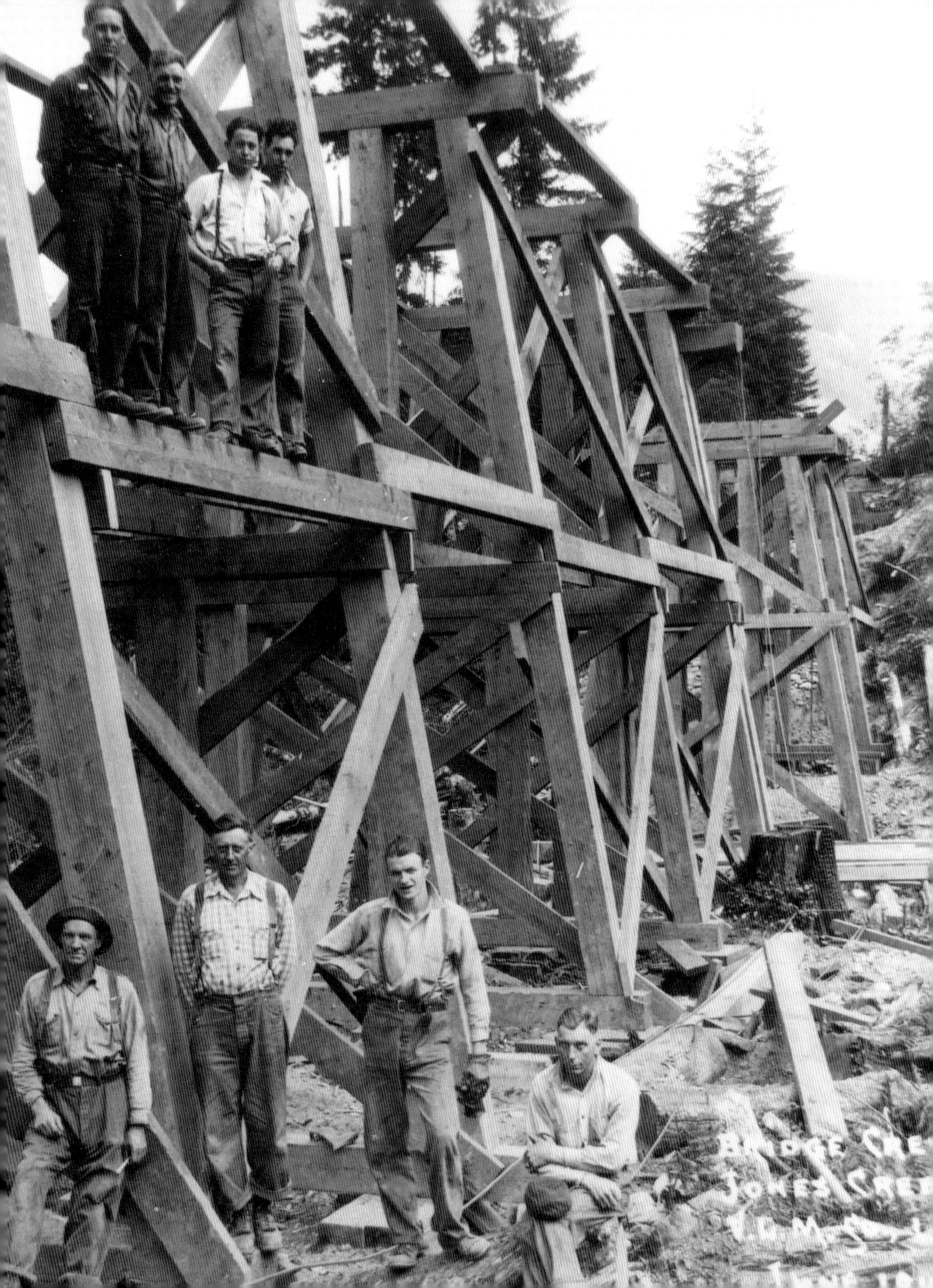
BRIDGE CRE
JONES CRE

CHRIS MEYLAND
THE BULL OF THE WOODS

The modern trend of things in this day and age is for a man to graduate from a university in order to become the big brains of an outfit; but when I first hit the big clearing, sixty years ago, the Bull of the Woods was usually a man who had graduated from a full rigged windjammer and had sailed the seven seas. Just such a man was Big Chris Meyland, the new Bull of the Woods who had just stumbled off the locie into two feet of snow. The wintry whiteness of the woods seemed to accentuate the blueness of his all-seeing eyes as he sized up Coburn's haywire outfit where he had that day arrived to make the round stuff roll. That was back in December 1919.

Young as I was, I could see we home-guards were in for some of the high-ball logging that we had heard went on in other camps. Big Chris tore into things right off the bat. We had just finished building a fore-and-aft road on a steep grade and the first log ran away like a greased pig and nearly wrecked the road donkey. Old Charlie Hosko, the skid road foreman, decided to put in a shear log and got into an argument with Big Chris, which ended in a rough-and-tumble fight with Charlie going down the road and us punks realizing just who was running the outfit.

Back in those days there were no crummies or speeders to take the crew out in, and we rode on the log cars with strict orders not to ride the locie. One cold January morning there was a foot of snow on the log cars, and when the locie stopped at the tank I sneaked up on the pilot of the locie to warm my hands. Chris stepped around and said, "Okay young fellow, better walk back to camp and get your time. That'll warm you up a little, eh?" But this being a home-guard outfit, I was hired back again when Chris cooled down.

In the spring of 1920 the Forest Branch decided to plant the logged-off land in red clover. It was thought this would keep down the fire hazard. Chris was skeptical at first but was all for it and hired a local farmer to do the planting along the sides of the railroad.

Now this camp of Coburn's had a little Shay locomotive which burned coal on the trips uphill and, as coal cost money and wood was free, it burned slab wood on the trips downhill. In order to save coal, the brass hats from the town office issued orders that as air brakes were not needed going uphill, coal could be saved by stopping

Opposite: Swinging logs from a cold deck with a skyline yarder, Copper Canyon, 1940. (UBC BC 1930/546/881)

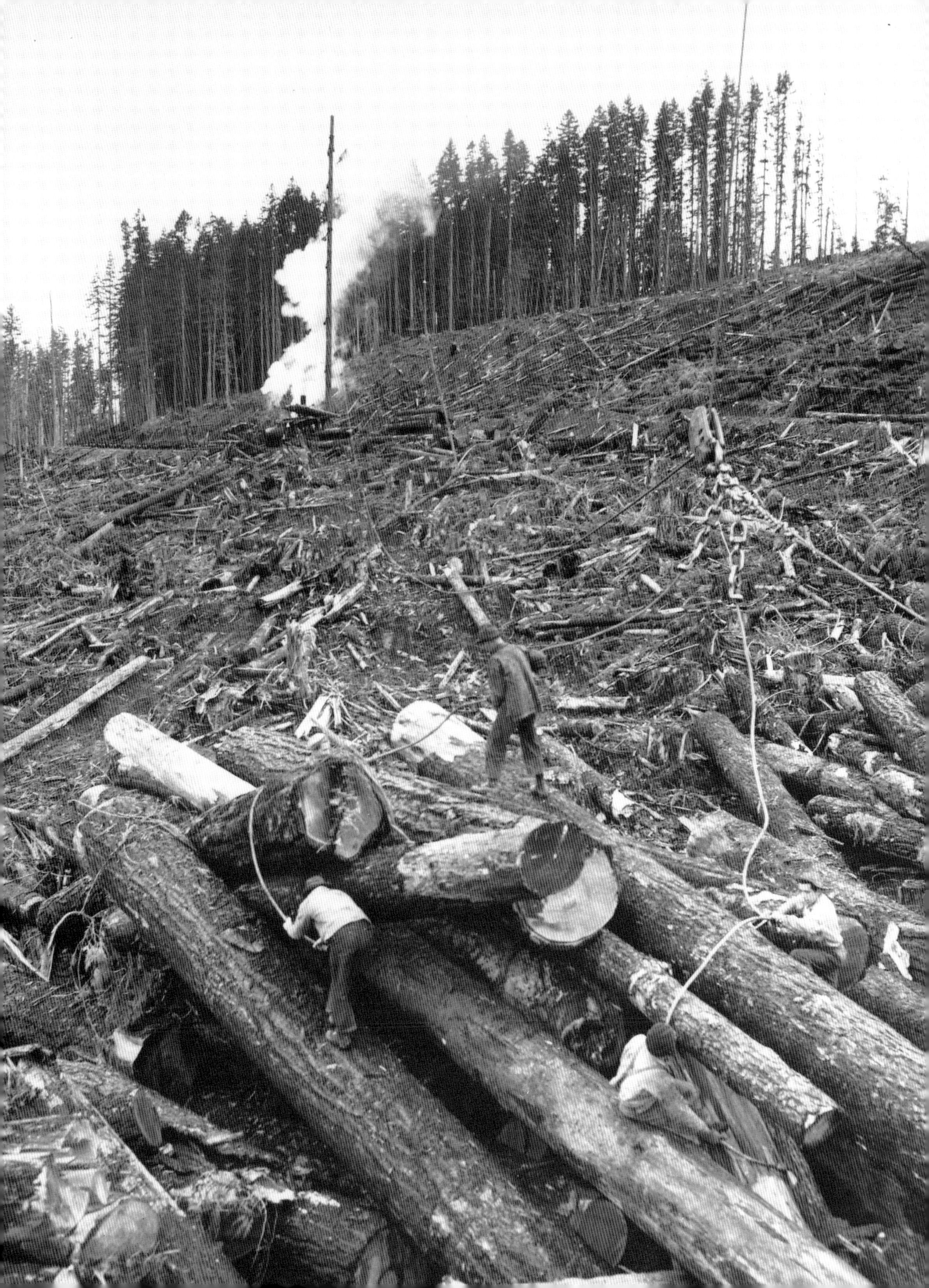

Moving a donkey at Bernard Timber's Orford Bay operation, 1926. (Bill Roozeboom photo, VPL 1412)

the air pump. Chris's answer was that there was far too much education around this outfit.

Next morning the locie arrived at the top of the hill all puffed out with the air pump shut off. A wood log was loaded and the locie signalled to drop it down a few car lengths so it could be dumped off in the road donkey's wood yard. With the pump stopped there were no air brakes and away she went, wood log and all, down the five percent grade. She gathered speed and the crew piled off into the weeds.

About a mile down the track the farmer was planting his seeds. Seeing the locie and cars dangling by, he waved at a phantom crew and kept on planting. He'd heard of Chris's highball logging and thought everything was as it ought to be. It wasn't long before Chris and the crew came galloping down the track and asked the farmer if he'd seen the locie.

"She went that-a-way," answered the farmer. "It shore is a fast outfit you run around here." The locie and cars piled up two miles down the grade and it took Chris about a week to pick up all the pieces. After that they ordered a car of coal and praised Westinghouse air brakes.

Among other things, Chris Meyland had a mania for pumps and water pipes. I've seen him stand for hours, fascinated, watching a hose run into the tank of a yarder. Strangely enough this peculiarity of his paid dividends one night back in 1920. A forest fire threatened to burn up the first cold-deck pile I'd ever seen. Three months of cold-decking would have gone up in smoke had Chris not jumped into the creek, retrieved a lost part of the steam pump and then put the pump together in the dark. Chris's knowledge of hydraulics that night saved a million feet of timber.

Big Chris Meyland was a past master at moving donkey. According to Seattle Red, who was an authority on such matters, Chris was the best donkey mover on the Pacific coast, and Red should know! However, Big Chris had a peculiarity unknown in the woods today. When the whistle blew for lunch he wouldn't quit, and because he was the Push no one else could quit until he did.

One time at Shawnigan Lake, Chris had just moved into a new setting and was hanging on the pass line between the jack guys and the bull block when the whistle blew for lunch. "Going up," hollered Chris as if he hadn't heard the whistle. Harry Todd, the donkey puncher, blew another dinner whistle but Chris wouldn't come down out of the spar tree.

With this the "fog-buster" hung a guyline shackle on the friction and they all went to lunch, leaving Chris up the spar to cool off and think over his shortcomings. When they came back he was still up there but when he did hit the ground a donkey puncher walked back to camp to interview the timekeeper.

Chris Meyland's last job was a labour of love and that was building a dam for the City of Nanaimo back in 1935. He died soon after that. In looking back now, I realize hard as Chris appeared in the old days, he was really a leader of men who had a kind of understanding for the underdog disguised beneath the rough exterior of a Bull of the Woods.

BIG JACK MILLIGAN

Big Jack Milligan, the old-time logger, threw his hat on the skid road and jumped on it with both feet. He'd hauled logs with bull teams, roaded logs with road donkeys, railroaded logs with Shay locomotives, and would have even tried to log with elephants if he'd been asked to do so. But this was the last straw and an insult to a logger's intelligence — whoever heard of hauling logs with *trucks*? They'd never work, Big Jack was convinced of that and he wasn't afraid to say so. This happened back in 1919 in Milligan's Camp at Jordan River when Orville Milligan, Jack's partner and brother, had bought a big rubber-tired truck and fitted it out with bunks to haul logs to the salt chuck. It was a new and revolutionary idea in the BC woods, and no wonder Big Jack was skeptical.

Big Jack Milligan was a real dyed-in-the-wool old-time logger. He was born in Quebec in 1871, and before he was fifteen he'd shot the rapids of the Quebec rivers on the big log drives with a peavey in his hands. He'd come west in 1895 and he and his brothers started logging with bull teams on the south end of Vancouver Island, in an outfit known as "Milligan's Camp." In 1908 the Milligans purchased the first steam donkey the Empire people ever built in Vancouver. It was a 10x12 open face with a cast-iron frame. Their next machine was an 11x15 Tacoma Roader that "roaded," or dragged logs on a skid road using a mile and a quarter of one-inch main line. This was known as the "Almighty Power" and Big Jack Milligan was mighty proud of her too. But this new fang-dangled affair — a logging truck with rubber tires — well, it's no wonder he headed for town to go on one of his customary two-week benders.

The first time I worked with Big Jack Milligan was in 1922, when he was tending hook on a track-side at Coburn's Camp, back of Nanaimo. Jack was the biggest and most powerful man I've ever known. He weighed 250 pounds and stood six feet six inches in the caulk boots. Like all famous characters of the BC woods, Jack's fame had preceded him and it wasn't long until we heard the story of how he'd come to be known as the best and loudest hollerer on the coast. No Iowa hog-caller had anything on Jack. There had been some controversy at the Milligan Home Camp between Jack and his brother as to who could holler the loudest. One day a test was made and Jack won. He could be heard over seven miles, so was declared and recognized as the champion of all hollerers on the coast.

One day in 1922 I had the pleasure of hearing one of his loudest shouts. We were yarding on a thousand-foot haul and the two track-side settings were about a quarter of a mile apart on the same back switch. Now it was one of those hot July days when the tree-boring beetles are on the wing everywhere in the woods. Perhaps I should explain: these beetles have a long stinger which they drive into the bark of a tree to deposit their eggs. Well, Big Jack sat dozing on a stump in the hot sun when a tree-boring beetle descended with bullet-like swiftness on one of his caulk boots and drove its marlin spike stinger clean through his boot and into his foot. Jack let such a holler out that the whistle punks on both machines shot in a signal and halted both yarders. Tough as Big Jack was he limped for a week after that freak accident, and killed every tree-boring beetle he could lay his hands on.

A ground-lead steam donkey working near Quatsino, 1900. (B.W. Leeson photo, VPL 14001)

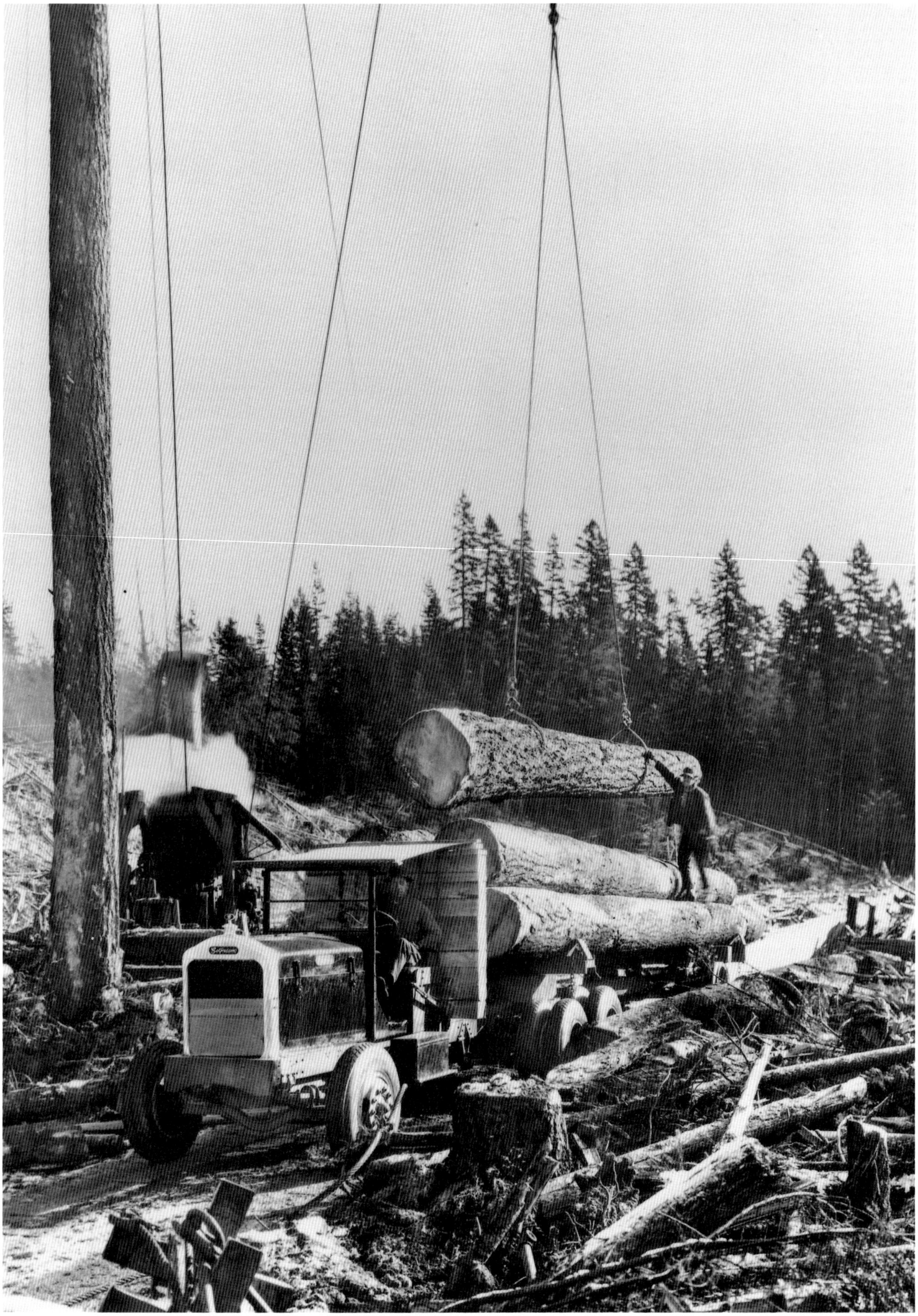

From time to time Big Jack went to work at his old home camp, but usually only stayed long enough to make a stake. He often told me those logging trucks got him down and it didn't seem like real logging as he knew it. Other camps were adopting them too, and it was beginning to look like they were here to stay. Road donkeys, like bull teams, had by now been discarded, and Big Jack didn't like it a bit.

The last time I saw him was in 1942. He was then too old to tend hook and had taken a job punking whistle at Port Renfrew. He'd slowed up a lot, for now he was seventy years of age, but he was the same old Jack as ever and could still make the canyons ring with his booming voice. He remarked on how things had changed—that he'd lived to see the day when trucks were replacing railroads and locomotives and it looked like they were going to be a success after all. Big Jack Milligan's career ended in Victoria where he died in 1944 at the age of seventy-three. And I've never heard to this day who *now* holds the BC hollering championship!

A mixed stand of Cathels & Sorenson Logging's timber near Port Renfrew, 1926. (Leonard Frank photo, VPL 5626)

Opposite: Loading a Gotfredson logging truck with a duplex loader at Commercial Lumber, near Haney, 1927. The driver is Herb Smith. (Leonard Frank photo, VPL 3697)

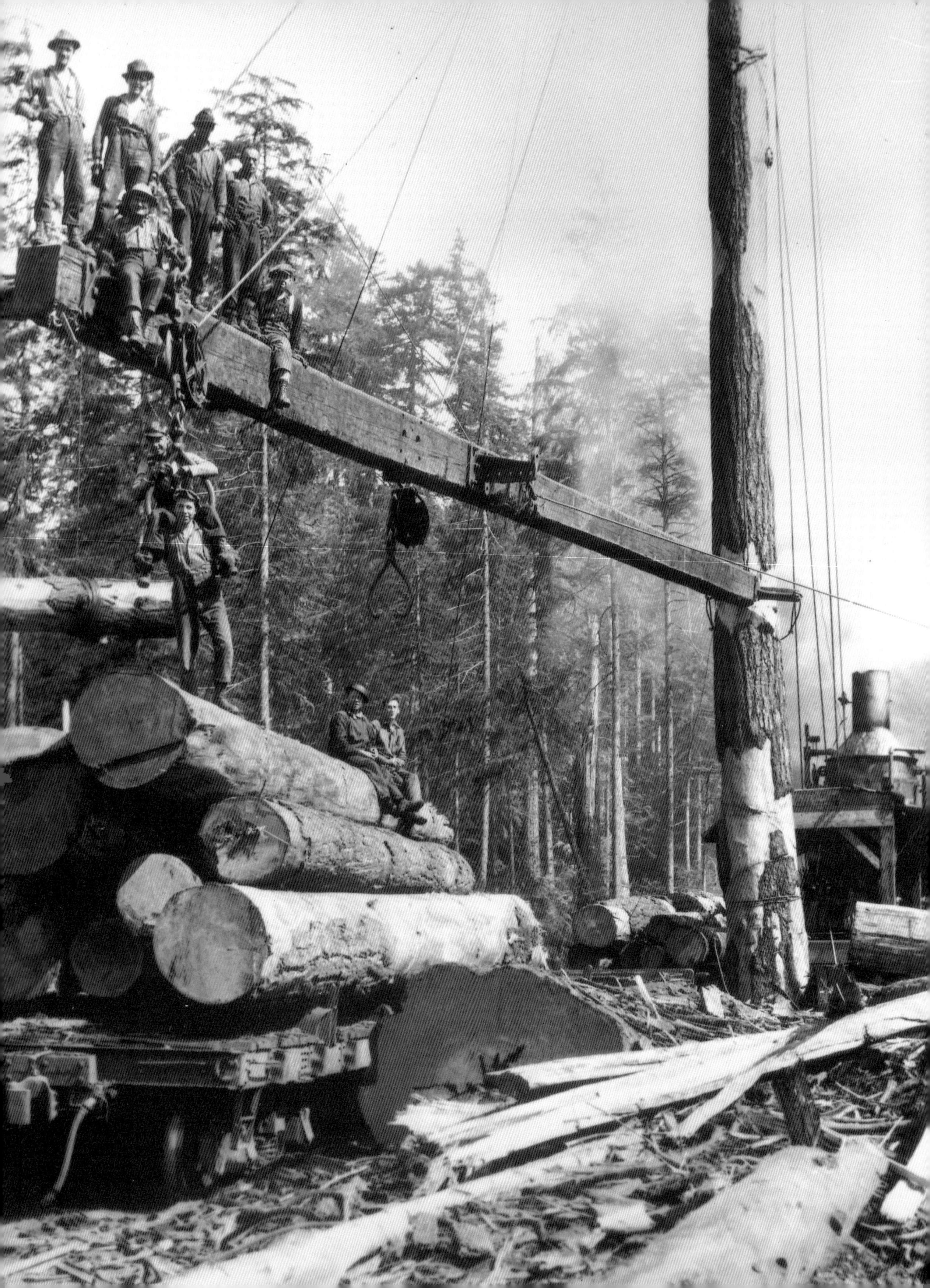

JESSIE JAMES
THE BIG TIME LOGGER

"You're just the man I'm looking for," said Jessie James to Rough House Pete, otherwise known as Pete Oleson, one foggy November afternoon on Vancouver's Skid Road. "How about getting your duds together and coming over to my new outfit at Cowichan Lake? We could sure use a good hooktender like you. Meet me at the boat and we'll go by way of Victoria tonight." That was Jessie James' way of hiring hooktenders for his new operation at Cowichan Lake. Jessie James, the sole owner of the Keystone Logging Company, had recently finished up his claim on the Lower Mainland, where back in 1929 the beer joints were still echoing with stories of the escapades of this swashbuckling renegade of all logging operators.

He had now acquired a tract of timber in the famous Cowichan Valley, and the new outfit was given the name of the James Logging Company. The new name merely made the outfit more "Jessie James" than ever, for a bar where beer was served to the crew was promptly set up right in camp. Jessie had even thought of hiring a dentist so there would be no need for a man to go to town to get his teeth fixed. All this put Jessie's camp in a category by itself, to the disgust of the more orthodox type of logging operators.

Well, Rough House met Jessie at the boat as arranged and landed in Victoria without any mishap. It was too late to start for camp that night so Jessie got a room and threw a party in honour of the newly acquired hooktender. Someone thought the beer was too warm, so Jessie filled the bathtub with bottles of beer and left the cold water running to cool it off. Pete was supposed to be watching it but he got too interested in one of the blondes. The labels from the beer bottles plugged the overflow of the bathtub and it wasn't long before the lower rooms in the hotel were flooded. Jessie offered to buy the joint but with the help of a couple of Victoria bobbies, the manager had Jessie and his party landed out on the street.

"We'll all go to camp!" James said, as he ordered his twelve-cylinder Marmon driven up to the curb. "We need a couple of flunkies in the dining room." So the girls climbed in and they all headed north over the then hazardous Malahat Drive. In those days the Cowichan Lake Road, from Duncan to Cowichan Lake, was famous for two things: one, its snakelike crookedness as it wound

Opposite: The loading crew with a MacLean boom at Victoria Lumber & Manufacturing Company's Camp 10, 1926. (VPL 1429)

its course between huge Douglas fir trees, and two, the speed at which Jessie James drove his big Marmon car over it.

It was breaking daylight when James and his party roared to a stop at the foot of Cowichan Lake and boarded his powerful speedboat, the *Stud Cat*, to begin their twenty-mile journey up the lake to Jessie's new camp. The throb of powerful motors put Pete to sleep while Jessie was explaining to the new flunkies they would arrive in camp just in time for breakfast where they would enjoy Jessie's favourite dish—rare beefsteaks. About a quarter of a mile from camp, the *Stud Cat* hit a deadhead and they all had to swim for it. The *Stud Cat* sank to the bottom of the lake in 200 feet of water. They were rescued and Jessie made the boast that he had at least arrived in camp with one hooktender and two new girl flunkies. The speedboat is still on the bottom of Cowichan Lake as far as I know.

Not long after that Jessie James made another trip to town and, while on one of his usual wing-dingers, demanded a rare beefsteak in one of Vancouver's best restaurants. While eating the steak he was listening to a good joke and laughed, and the steak lodged in his throat and choked him to death. Thus ended the career of BC's famous Jessie James. He had lived dangerously and had cheated death at breakneck speeds over crooked roads, braved the icy waters of Cowichan Lake, but met his Waterloo eating a beefsteak in a city restaurant—a pretty feeble ending for so tough a logging operator as Jessie James was supposed to be.

Opposite: A wooden skyline spar at a Victoria Lumber Company show near Chemainus in the 1940s. A MacLean boom is being used to load the cars. (Jack Cash photo, UBC BC 1930/547)

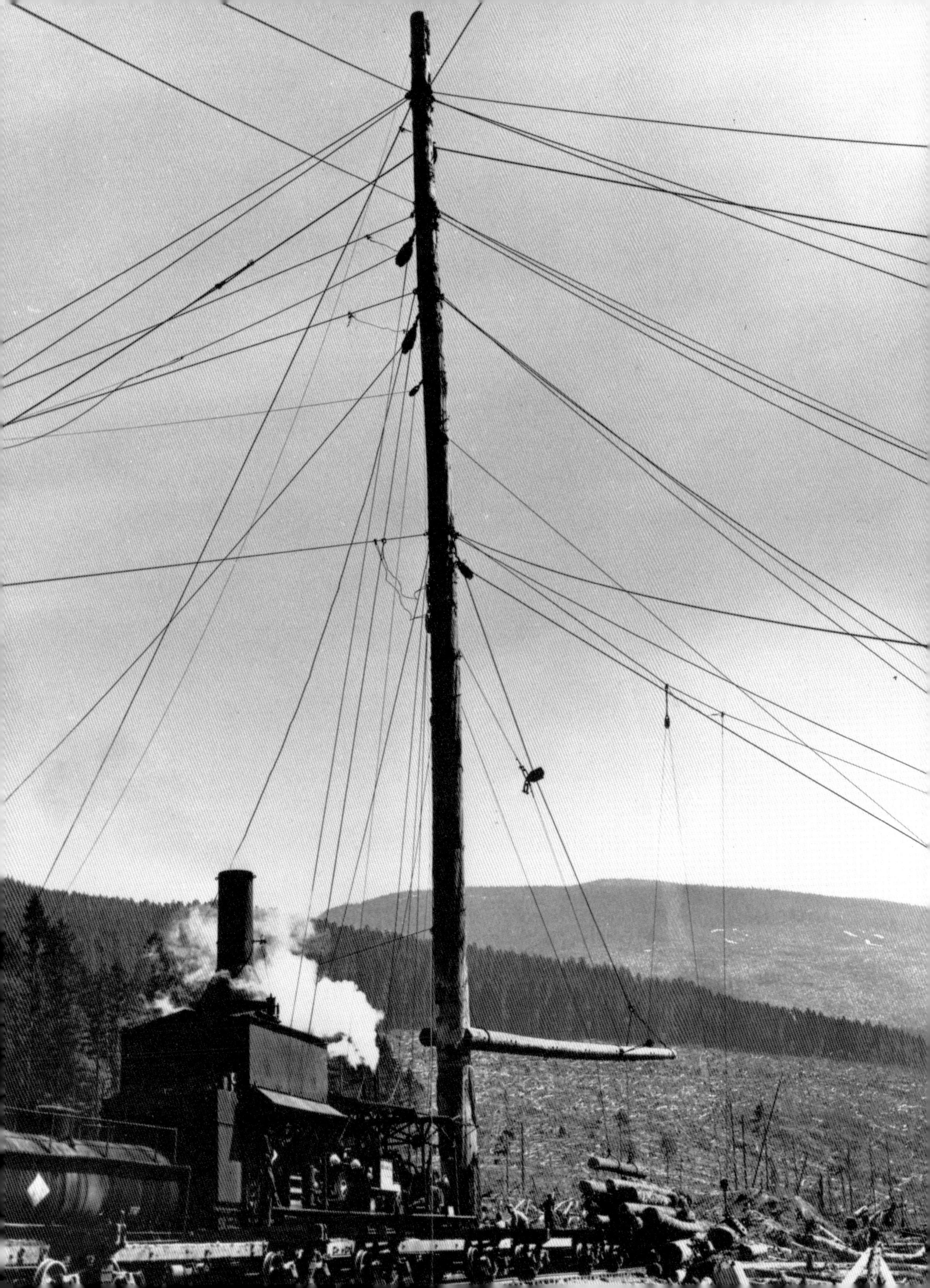

SEATTLE RED

Seattle Red, the fiery-tempered, rip-snorting rigging slinger, threw his gloves in the air and didn't wait for them to come down. Even the whistle punk was haywire. It was bad enough to log with haywire but when the crew was both haywire and home-guard too, it was time to pull the pin.

He didn't even wait for his time but headed across country and hit for town.

This happened back in 1928 when Seattle Red hired out to pull rigging at Hill 60 near Cowichan Lake. After working a few hours with the home-guard crew, he left them wondering where he had disappeared to. They searched the road, thinking they had hauled logs over him. What they didn't know was that this was Seattle Red's way of quitting a home-guard outfit.

Every old-time logger who's worked on the coast will remember Seattle Red. He'd worked in about every camp of any importance from Sitka spruce shows in Alaska to the sugar pine camps of California. Contrary to popular opinion, Red wasn't born in Seattle. He was born in South Africa where his parents lived, and where his father was a building contractor about the end of the Boer War. From there the family moved to Vancouver Island where they bought a ranch near Nanaimo, BC. It was on this ranch that Danny, later known as Seattle Red, was introduced to a falling saw, a springboard and hammer and wedges before he was ten years old. And at that tender age Red learned to holler "Timber-r-r-r!" with the best of 'em. I can speak with authority on this because I happened to be Danny's little brother.

When Danny was sixteen and had finished school, he got a job with the CPR at Wellington, BC as a car-knocker's helper, but he could hear the donkey whistles and the sounds of ground-lead logging on Mount Benson only a mile or so away, and so he quit the CPR and got a job out there punking whistles in the woods and he never looked back.

It was in 1926 that Seattle Red got his name. He was by then working on the rigging at the Abnernethy Lougheed camp, out in the Fraser Valley, when he decided to cross the border and work in the State of Washington. It wasn't easy to get past the US immigration officers so Red ducked across the border during a thunderstorm and cached his pack. Then he headed for the railway tracks and walked north until he came to the border. He was stopped by a Canadian immigration officer, who asked him where he was from. Red replied that he was from Seattle and was heading for Vancouver, BC.

"Well," said the immigration man, flashing his silver buttons, "you'd better head right back to where you came from," and he hurried him back to the United States, right past the US Immigration. So Red hiked back, got his things and thumbed a ride to Seattle, where he promptly hired out to a US logging camp south of Seattle.

During the year of 1926, Seattle Red worked for no fewer than thirty-five logging companies in Washington, Oregon and California. He stayed only long enough in some of the camps to wash his

The cookhouse at Abernethy & Lougheed Logging's float camp on Alouette Lake, 1926. (VPL 1477)

The rigging crew at Bloedel, Stewart & Welch's Menzies Bay camp, 1926. (VPL 1503)

clothes and get a good feed. By the end of the year the hiring officers in Seattle wouldn't hire him out any more – said he was a tramp – so he rode the roof of a boxcar into Portland. There he pawned his watch (a big seventeen-jewel Hamilton) to pay his fare out to a camp in sugar pine country in California. When he'd saved a stake he headed back into Seattle again, but being a true logger he was soon heading back for the sticks.

Now besides being a high-ball logger, Red was a philosopher in his own way. He decided that as he was bound to spend his money anyway, he was going to have some fun out of spending it. He'd always had a passion for automobiles – big automobiles – so he saved another stake and bought a huge, flashy Marmon roadster. Then he spent his money on gasoline instead of whiskey.

The first time I actually worked with Red, outside of working at home on the ranch with him, was in 1920. We were setting chokers in snow knee-high to a giraffe, and it was then I learned to appreciate his true sense of humour. The outfit had a number of haulback blocks made by Young Iron Works. These were known as "Young" blocks. Red was packing them through the brush on his back and he dryly remarked to the hooker that if these were young blocks he would hate like hell to pack them around when they were fully grown. Another time at Cowichan Lake, Red was put second loading on a high-ball track-side. That night he walked into the office and asked the foreman if he could have a horse collar.

"A horse collar?" asked the puzzled man of authority. "Whatever in the world would you want that for?"

Moving a gas-powered cold deck donkey at Rock Bay, 1934. (BCARS 73701)

Yarding at Victoria Lumber's Copper Canyon show, 1940. (Jack Cash photo, UBC BC 1930/546/893)

"By the lightning blue Lord Harry," Red exploded, "if I have to work like a horse I might as well dress up and look like a real one."

I visited with Red over on the Island back in 1951. He was tending hook back of Ladysmith. I saw quite a bit of him. He was a great help in the testing of the whistle I was developing for the new diesel locomotive that CPR was trying out on the E&N Railway. The new diesel sounded like a sick moose. In fact, up north on the CNR the bull moose in the rutting season challenged the honking diesels and a lot of bull moose were killed. When it came out, the new whistle had no moose appeal. My brother Seattle Red was elated when the first new whistle arrived on CP Engine No. 8001, and here's the letter he wrote to me August 23, 1951.

Ladysmith, B.C.
Thursday, August 23/51

Dear Bob:

Dawn, on a placid misty island dotted bay on Vancouver Island, time 5:40 a.m. To be gently, but certainly awakened by the rich, melodious tone of an M3, and to realize that at long last they are here, is most certainly like waking up in Paradise with an angel on each shoulder, and then, as I stuck my sleepy head through the open window the "turn around" rolled in. There, in the dim dawn light, I could just make it out, yes sir, that was her, with engine 8001, and there, snugged down in front of her cab, was most surely to me, the crowning cry of two and one half years eager despair.

If anyone had spoke to me at that moment I would have been unable to answer them, as a lump came in my throat, and, as I lit a cigarette, from my eye I dried a tear of contentment. It seemed the irony of fate, the same engine that pioneered our earliest endeavour on that cold morning of Jan. 1949.

Hoping to see you one of these days.

Dan

Seattle Red became a home-guard at Ladysmith and he stayed with Crown Zellerbach and logged at Nanaimo Lakes. He gradually slowed down, and his last job was punking whistle until he took retirement. It was a sad day for all of us when he passed on in 1968 at the age of sixty-seven. He was interred at Ladysmith within hearing distance of "that whistle in the night."

MATT HEMMINGSEN

Matt Hemmingsen strode over the fifteen million feet of logs jammed in the Tsolum River at Courtenay. The river was plumb full of Douglas fir from the beach clear up to Headquarters – a distance of about six miles – and all attempts to drive them to tidewater had failed. It was a sweet mess, and no wonder the coast loggers thought Old Humbird, the big shot at Chemainus, was out of his mind when he'd ordered the river filled with logs and forbade the building of a skid road along its banks. This happened back in 1907 and Matt Hemmingsen was just the man to clean up the mess – that's why he'd been sent out from Wisconsin.

Now, back in Wisconsin Matt had worked on the river drives for a good few years. He started in the woods back there for the Humbirds at the age of fourteen, and before that he was brought up on a stump ranch. He started in the woods in 1889 and his first job was flunkying; but after ten years around the Wisconsin woods he could drive anything from a bull team to a roaring river. Breakfast was served in those camps at a quarter to four in the morning. Matt was judged as one of the most likely young loggers to succeed as he could eat as many beefsteaks and hotcakes as the best of them. His pay was sixteen dollars a month and board and he worked sixteen hours a day. By 1902 he'd worked his way up to foreman and was in charge of an outfit.

One time, on the spring drive, all the foremen were eating lunch on a little island in a Wisconsin river when Humbird, the boss, came out to camp to look over the outfit.

"Which one of you fellows is young Hemmingsen?" asked the big timber king.

"I am," replied young Matt from the circle of foremen.

"Well," bellowed Humbird, "you're getting so big I hardly knew you – but here's what I want to tell you. You logged the cheapest logs ever to be driven down White River. Now, when this drive is finished, Hemmingsen, I want you to go out West to a place called British Columbia and be logging superintendent at Chemainus. They log a lot differently out there than they do here. Sit on a stump for two or three weeks until you learn how it's done, and then fly at her. But get logs – that's all I want."

Matt Hemmingsen, second from right, with a Washington ground-lead yarder operating west of Chemainus in 1906. (BCARS 53598)

And so Matt Hemmingsen landed on Vancouver Island on June 2, 1906, to find the Tsolum River plugged with fifteen million feet of timber. He sized up the situation with a practised eye. Then he carefully blasted the rock bends out of the river and, with the fall floods, the drive started. Matt rode the last big blue butt right into Courtenay. The booming ground was filled with logs six tiers deep and a million and a half feet went out to sea.

The Vancouver Island loggers were still raving about this Wisconsin logger when he rigged the first high-lead tree back in 1910. They were yarding out of a big swamp and all the logs bogged down. They had no belt and spurs, but a hooktender by the name of Anderson climbed a big hemlock by driving in railroad spikes. They didn't top the tree but hung three guylines and the old ground-lead bull block and started in to do the first high-lead yarding. The system was a great success and from that humble beginning the great high-lead system of logging was developed. Six years later Matt made a trip down to the Grays Harbour country, where they were still ground-lead logging, and set up the new system for them.

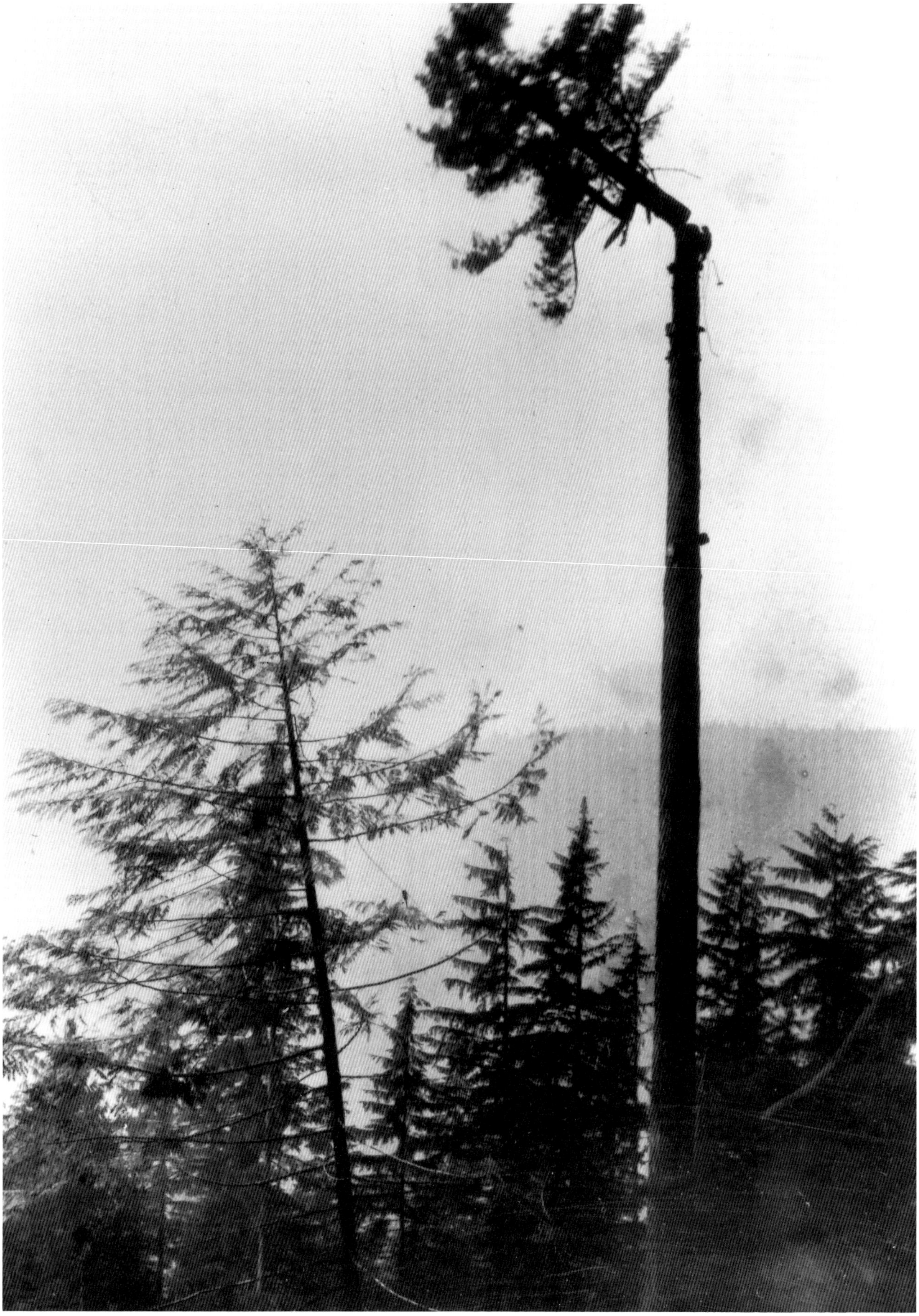

It was in 1912 that Hemmingsen went into the Cowichan Lake Valley to open up that famous Douglas fir stand of timber, and it was at Wardroper Bay, on Cowichan Lake, in 1914 that Matt rigged up the first skyline ever to be used in the BC woods. Of course he'd used skidders at Chemainus as early as 1910, but his new skyline system was soon universally adopted. It was at Cowichan Lake, too, that Matt had such famous characters working for him as Rough House Pete, Eight-Day Wilson and Johnnie-on-the-Spot, and he occasionally had a run-in with the notorious boss logger Jessie James.

By 1915 Matt decided he'd go logging for himself. He turned down an offer of $20,000 a year to run the old Empire Lumber Company, which later turned out to be the big Youbou outfit of today. He took a contract to log for Palmer, of the Victoria Lumber & Manufacturing Company, but when the Chemainus mill burned down in 1923 Matt was left with thirty million feet of logs in the water and no mill to cut them.

The loading crew at Victoria Lumber & Manufacturing Company's Camp 10 at Cowichan Lake, 1926. (VPL 1433)

Opposite: A high rigger tops a Douglas fir to use as a spar tree at Abernethy & Lougheed's Camp 3 on Alouette Lake, 1926. (VPL 1488)

Cathels & Sorenson, at their Port Renfrew camp, used the booms along the shore to lower logs into Davis rafts, 1926. (VPL 5611)

By now Matt figured Cowichan Lake was hoodooed. He'd never had any luck up there. Why bother with logging? He'd always had a hankering for mining and for years had been grubstaking prospectors. Now it looked like he'd hit on a good thing up at Revelstoke. So in 1930 he quit logging and went to Revelstoke to cash in on the $75,000 he'd spent developing the ground. But by 1935 the pioneer logger found out that all was not gold that glittered. He says he didn't exactly go broke – he did get nine dollars back out of the $75,000 investment. He even tried to develop a marble quarry but that, too, went the way of most mining ventures. Finally Matt decided he'd better go back to logging and found himself looking for a bit of timber.

The bit of timber turned out to be at Port Renfrew, where he founded the Hemmingsen-Cameron Timber Company in 1938. This was a bold venture on Matt's part for his new company had

In the Queen Charlotte Islands logs were rolled onto Davis rafts with the aid of a small tug, 1918. (UBC BC 1456/62)

An assembled Davis raft of airplane spruce ready for towing across Hecate Strait, 1918. (UBC BC 1456/62)

The Cathels & Sorenson Logging Company camp at Port Renfrew, 1926. (VPL 1436)

taken over the assets of the old Cathels & Sorenson Company, which was, incidentally, the boneyard of just about all logging outfits on this continent. There were thirty-eight steam donkeys lying around the woods with alders growing up between the drums. Every machine was falling to pieces. The railroad was as crooked as a dog's hind leg. To put the cap of gloom on the whole picture there was an old Climax locomotive. It, too, was haywire! Well, Matt was used to haywire by now and he sailed right into the tangle of junk with a new lease on life, for the timber was good and that meant a whole lot. It took eight years of hard slugging. It wasn't easy fighting the seas of Vancouver Island's west coast with Davis rafts, building new camps and railroads, but he finally got the outfit on its feet and sold out to the BC Forest Products in 1946.

I called on my old friend Matt Hemmingsen some years ago in Victoria where he lived in retirement with his wife and grandchildren. I'd always thought somehow he was a Norwegian, but he told me he was born in Ashland County, Wisconsin, in 1876 and that his full name was Mathias Hemmingsen. When I told him he was a lucky man, he replied in his usual gruff manner:

"You know, Swanson, the only lucky thing that ever happened to me was the day I married my wife at Courtenay, back in 1910, and after thirty-six years together through thick and thin I realize I'm a pretty lucky old fellow at that."

GEORGE HANNEY AND THE FLYING DUTCHMAN

Most of the pioneer loggers on the BC coast are famous for the daylight they have let into the swamps. But George Hanney is a pioneer who was always content to wait for daylight to creep in through the cedars as he sat by the edge of a swamp with a rifle across his knee. His name will always be as much a part of the timbered valleys on Vancouver Island as the deer and the bears that roam through them. He was a hunter and a trapper but occasionally he worked in his beloved woods as a logger.

George Hanney was born in England in 1885. By the time he was fifteen he had landed on Vancouver Island as a ship's apprentice. In 1902, at seventeen, he hired out as a ship's pilot to navigate steamers and windjammers into the harbours of Ladysmith and Nanaimo for cargoes of coal and lumber.

Jack McNeil (arms folded) with crew using ground-lead steam yarder and line horse at Stocking Lake, near Ladysmith, in 1902. (BCARS 53607)

Victoria Lumber & Manufacturing crew with early one-drum steam yarder near Ladysmith, 1903. (UBC BC 1930/142)

It was in 1904 that he hired as sniper at old Camp 6, back of Ladysmith, for the Victoria Lumber & Manufacturing Company. E.J. Palmer was Bull of the Woods and Hanney worked partners with Parker Williams and Charlie McGarrigle. On this job they logged with a spool donkey and line horse and did their roading with a Climax locie that ran on rails spiked to the cross skids of the skid road, and they dragged the logs for miles between the rails to tidewater.

One day George Hanney was working down at the boom when Mr. Palmer fell in the salt chuck. After they fished the big shot out of the drink, Hanney asked him how the logs looked from the underneath side. Palmer fired him! After that he worked in the mines at Extension but had a trap line in the mountains where he spent most of his time.

George Hanney's fame reached its peak after he joined the BC Police Force in 1908. Many of his experiences on that job are today legends over on Vancouver Island. For instance, there was the time the notorious "Flying Dutchman" escaped from the US authorities and hid out at Union Bay. The Chief of Police sent George Hanney, and Constables Ross and Westaway, to apprehend the criminal. It was suspected the Flying Dutchman and his gang had a hideout over on Denman Island and at night rowed over to Union Bay for food. The gang got it by raiding the big false-fronted store that stands today near the Nelson Hotel at Union Bay.

With a little detective work Hanney and his men cornered the Flying Dutchman and his partner in the store but the hoodlums shot it out with the police. Constable Westaway was killed but Ross handcuffed himself to the Flying Dutchman after a rough-and-tumble battle. The other gangster got away, and it was here that George Hanney's trapping ability came into play. He tracked him for weeks and finally discovered the gang's hideout on Denman Island.

A big log lay in front of the bandits' cabin where Hanney concealed himself and waited. He could see there was another man in the cabin but the idea was to catch them both. Finally, after a long wait, the wanted man sneaked up one side of the log, and George followed him up on the other side, keeping them both covered. At the point of a .45 Colt, Hanney forced them into a rowboat and made them

Logging crew at the bunkhouse of Camp 5, Victoria Lumber & Manufacturing Company, 1903. (UBC BC 1930/144)

row him over to Union Bay where he clapped them both in jail. One bandit turned King's evidence and the Flying Dutchman got his neck stretched in the Old Nanaimo jail.

Another time George Hanney was sent out to see if there was any pit lamping being done on Blackjack Mountain. He was going along the trail with a pit lamp on his head when a hunter took a shot at him and blew the lamp right off his head. It was a close shave, but the boys who used rifles were good shots over on the Island in those days.

Not long after that Chief Justice Morrison told him to be very careful when he went out in the woods as he might be mistaken for a deer. To which Hanney replied: "Did you ever see a deer with a lamp on its head?"

Along around 1915 the call of the wild was stronger than the call of the police force. So George quit the force and went back to his trapping. He had a trapline and a cabin in back of Cameron Lake where he spent the next ten years of his life.

By 1930 logging operations had penetrated so far into the timber that Old George had to give up his trapping for a living that was more profitable. During his many years in the uncharted timber he had come across many mineral deposits and coal outcroppings. So it was small wonder he was appointed to make a survey and prospect for coal in the Alberni Valley.

When I interviewed the old pioneer over in Alberni, he showed me his report on the coal beds in that district. At one place he had located a thirteen-foot seam which he hoped one day would supply coal for the fast-growing towns on the west coast.

And there I left him happy in his labour, the true spirit of a pioneer. A never-flagging faith in the natural resources of British Columbia drove him ever forward.

RED MORRISON

Red Morrison waited for a twenty-foot tide to float his first logs to Seaford's new sawmill at Jedway on the Queen Charlotte Islands. That was back in 1908. Red had been sent up from Vancouver with a crew of ten men to handlog for the new mill – the first commercial logging venture on the Queen Charlottes.

To Red Morrison the mammoth Sitka spruce was a new and fascinating sight, but no more so than the antics of the Jedway citizens as they capered around the streets beyond the pale of the law. When the police finally got around to arriving on the scene the first thing they did was to build a log jail – and the first man to occupy it was one of Red's loggers, a fellow by the name of Jack McNeff.

Now, the cops weren't so hard-boiled in those days and they knew a logger could be trusted. They gave Jack a rowboat so he could row over to the cookhouse for his meals whenever the cook rang the bell. After his meals he would row back and hang around the vicinity of the jail. This kept up until his sentence had been served and he went back to work for Red. You see, Morrison had told the cops good loggers had to be treated right, and after fifty years around the Pacific coast woods this was still the motto of this famous camp foreman.

Red was born in Olympia, Washington in 1885. As a boy he hung around the livery stables where the boss loggers kept their driving teams. In those days a logging operator's success was judged by the driving team he sashayed into town with.

One day the big tycoon logger John Jemerson drove into the stable with his prize pair of greys. While Red was admiring them, the big shot asked him if he thought he could drive the team out to camp. He had to go by train to Spokane. Red replied if they were whip-broken, and if he knew where the camp was, he and his partner would drive them out there.

"You don't need to know where the camp is," said the Brains. "Give the horses their heads and spare the bit and they'll get you there. And Red, give this note to my foreman and he'll put you to work."

And so at the age of thirteen Red Morrison was steered into his first logging camp with good horse sense. The stable boss was

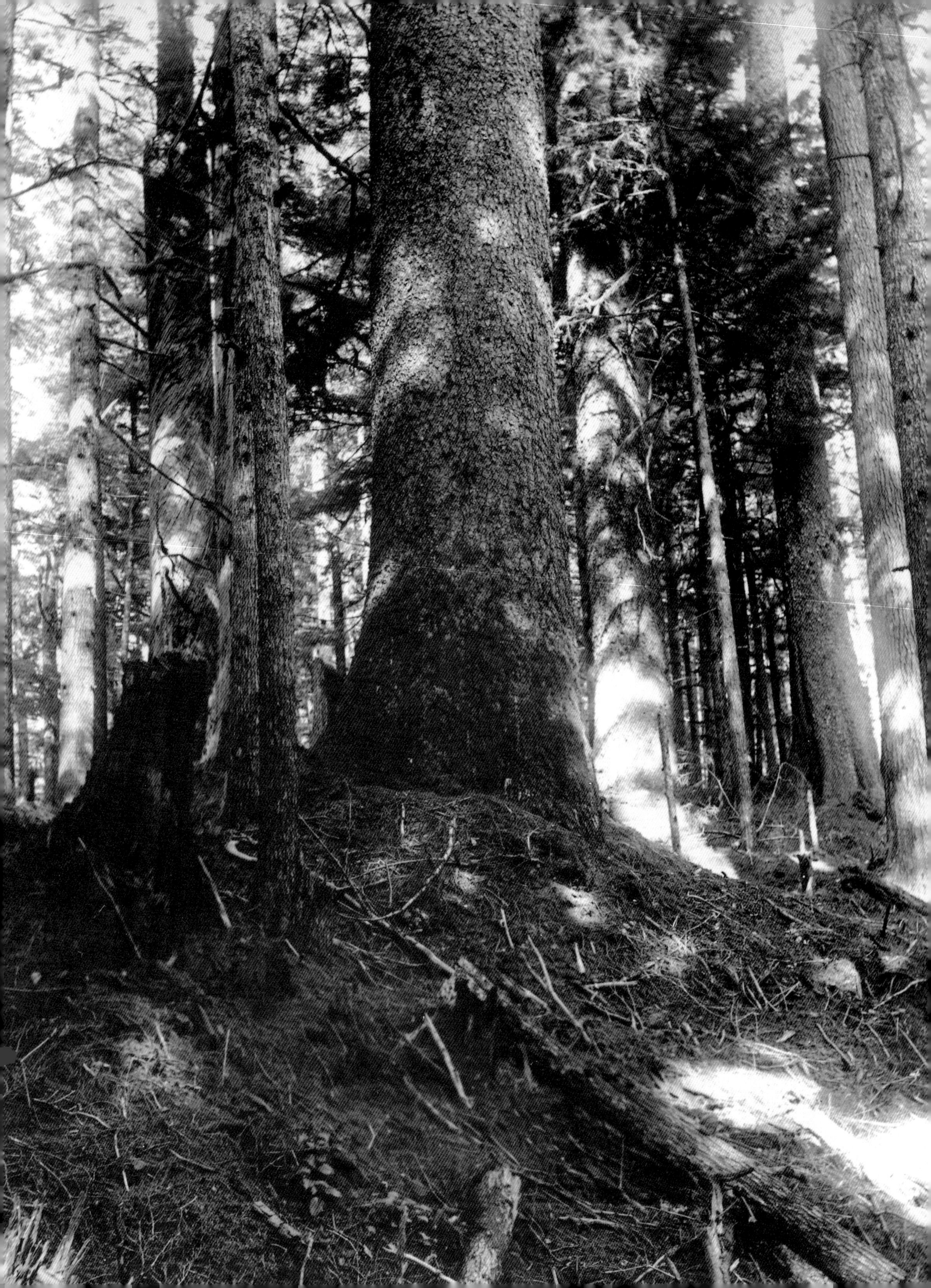

Horse team hauling logs on a sled along a hewn log trestle (location and date unknown). (VPL 30459)

furious when they arrived and Red was given a job as flunkey until the big boss arrived back in camp. That was back in 1898. All the big outfits were horse logging in those days and Red believed that a horse responded to kindness and understanding the same as a human being. His next promotion was driving a line horse that rode into camp on a flatcar with the crew every night.

It was in the fall of 1902 that Red Morrison landed in British Columbia. His first job was up at Dirty Face Jones's camp at Elk Bay. The camp was closing for the winter when Red arrived, and already the horses were on their way to town, as Jones had bought an 11x15 Vancouver steam donkey. Red was broke, so he asked for the job as camp watchman for the winter. Gradually, a few at a time, the loggers went off to town until Red was left in camp alone with not a horse, nor even a dog for a companion.

It was then that the solitude of the forest settled like a gigantic curtain around him. Before Christmas it began to freeze and he was kept busy keeping the pipes from freezing up. Then a flurry of fine powdery snow began to drift in the air and the ground turned white. The snowflakes got bigger, until by New Year's Eve there was five feet of snow. That night from out of the wilderness there came about

Opposite: A typical stand of Queen Charlotte Islands spruce, 1918. (UBC BC 1456/62)

Queen Charlotte Islands spruce forest in winter, 1918. (VPL 3740)

forty cats of every variety. One thing they all had in common was they were all very hungry — and noisy. Red had no gun to salute the New Year so it was done with genuine catcalls. Red fed them all. Not long after this a pack of wild Indian dogs came to camp, and they were fed also. Red made a pet out of one little fellow and called him "Sharkey." The next to be driven in by hunger was a herd of deer. These Red housed and fed in the old barn.

A few nights later a family of hungry cougars arrived; with no gun around, Red's menagerie was at their mercy. First the cougars tuned up on all the cats. Then all the dogs vanished except little Sharkey. Then the deer were their legal prey; but Red had them boarded up in the old barn. At nights the cougars could be heard prowling around camp and scratching under Red's bunkhouse. The pet dog would jump into bed with its master, and when daylight would break each morning Red and his dog would go around to the barn to feed the deer and repair the badly clawed barricade.

A Queen Charlotte Islands sawmill, 1918. (UBC BC 1456/62)

    One day along toward spring Dirty Face Jones arrived in camp and from the bunkhouse door shot four cougars without lowering his gun. Red breathed easy once more and as Jones, in spite of his traditionally dirty face, was a pretty good cook, he cooked a meal and rang the cookhouse bell. To Red Morrison the sound of that bell was the opening bars of "The Rustle of Spring." But to the pet dog, Sharkey, it was a death knell. Just as the bell finished ringing, a cougar rushed in and killed the little pet. Thus ended Red's long winter vigil at Elk Bay back in 1903.

    The first time I worked for Red Morrison was over on the Island in 1922. He was by then foreman of a big high-lead show. I was punking whistle but had an uncontrollable desire to oil up the donkeys at noon. Well, I ate my lunch out in the woods after the day Red caught up with me, and I still have a horror of oil cans!

HARRY TODD
KING OF DONKEY PUNCHERS

The first time I saw Harry Todd juggling the levers of a 10x12 Washington steam donkey, I was spellbound. I realized I was in the presence of a king—the king of all donkey punchers. It was a thrill to watch him manipulate that whirling, rattling mass of machinery while he rolled a cigarette at the same time, and then, as he plunked the turn at the foot of the spar tree, light the cigarette before the chaser could get the jewellery unbuttoned. I had the distinct honour of firing for this master. He taught me the subtle art of coaxing big blue butts out of timbered canyons in royal style.

Harry Todd got his steam ticket in 1904. Its number is 943 on the Engineer's Register of British Columbia. At that time he was engineer in a circus, where he met an old-time hooktender who coaxed him into quitting the circus and going into the woods to punch his first steam donkey. It was a vertical spool machine with a line horse and a spooltender. One day Todd got the heel of his boot in the gears and stalled her. Gears were not guarded in those days. The push asked him if he thought he was still running a merry-go-round.

Before his escapades as a logger, Todd sailed with the sealing expeditions to the Bering Sea. He was out in a canoe at one time, off the Pribilof Islands, when a fog came up and he got lost. He drifted for days in the fog and was finally picked up by a German windjammer and put ashore in San Francisco. In return for his kindness, Todd gave the German captain the canoe. But its price was deducted from his pay when he reached his home port of Victoria. So he "borrowed" the vessel's ten-gauge shotgun and quit sealing.

I think Harry Todd was the best donkey doctor ever to wield a Stilson wrench. He was a genius at tricky jobs without a helper. One Sunday he was monkey wrenching his machine and the boys noticed he didn't come in for supper. A man was sent out to see what had happened and it was found that he had rigged up a "Spanish windlass" with a piece of strawline to spring a steam pipe. He had got his hand fouled and was trapped. After that he usually took the fireman out with him on Sundays.

We were yarding out of a deep canyon back in 1922 on a 1200-foot

Opposite: Logging spruce in the Queen Charlotte Islands with a steam donkey and a wooden spar, 1918. (UBC BC 1456/62)

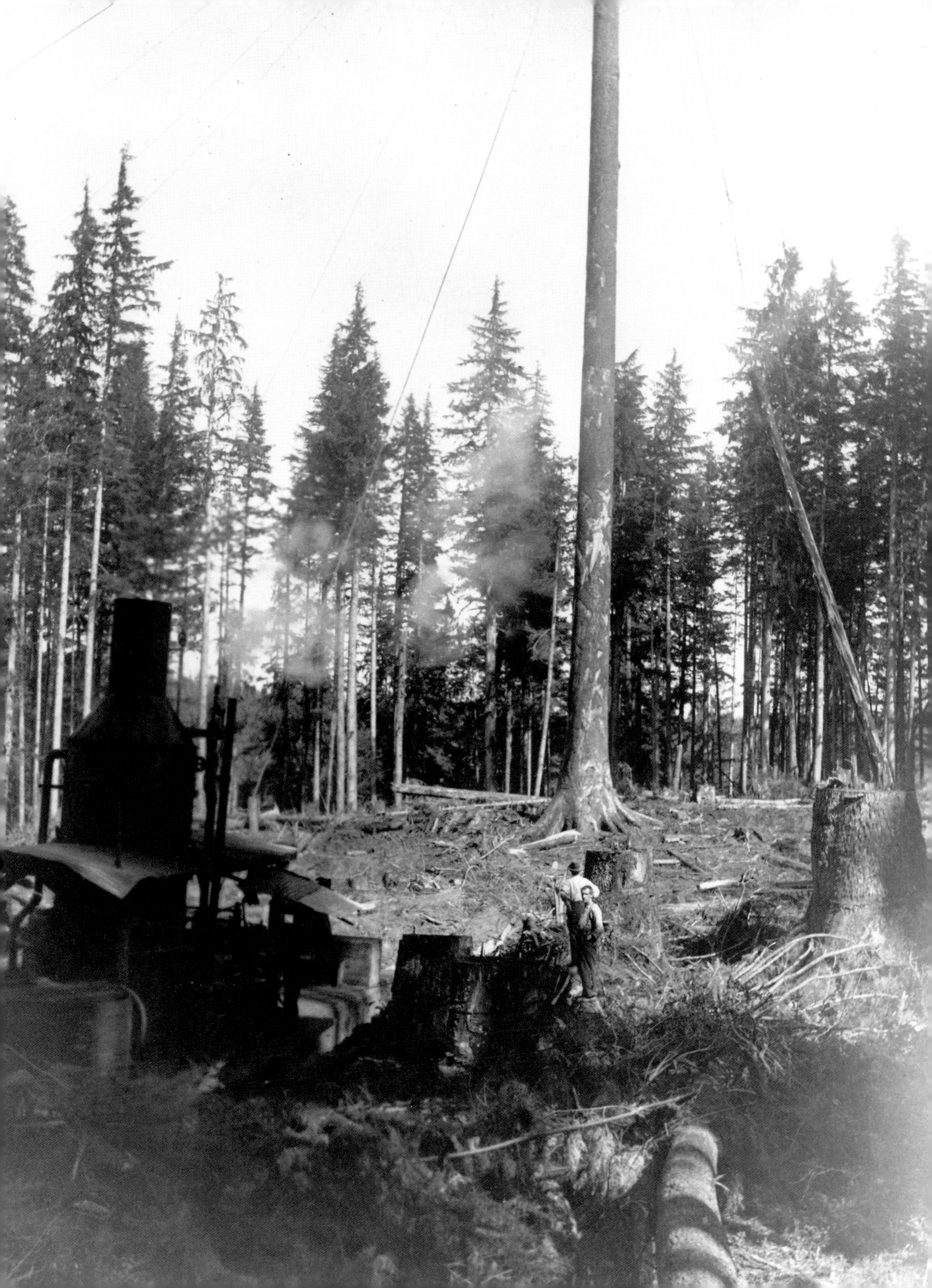

Comox Logging & Railway Company yarder at Tyee Lake, 1914 (CDM)

haul and were hung up most of the time. The Push remarked one day that he wished we had a bigger donkey. "This one's plenty big enough," Todd butted in. "Just watch me and I'll show you how it's done." He then whistled for the hooker and told him to put on a bridle hold, using two chokers on the top end of the big 70-foot stick. Then he told the fireman to fill the firebox with pitchwood and kindling. When he got a whistle to go ahead, he sent the fireman up on the roof to hold down the safety valve with a crowbar.

Black smoke rolled from the stack for about ten minutes. Then Todd tooted a very high-pitched whistle, took up the slack, tilted the throttle wide-open and held her there. Red-hot cinders showered around the poor fireman up on the roof. The turn started and kept coming. When he landed the big stick at the spar tree the fireman let go – and so did the pop valve, for the boiler still had a full head of fog even after that mighty burst of power.

Todd never said how much steam he had but casually remarked: "Never saw a boiler blow up yet; and I never saw a log that couldn't be hauled, either."

That was typical of Harry Todd at his best.

Harry joined the Forestry Corps and went over to Scotland to do a bit of high-ball logging in 1940. When he returned in 1944 I again had the pleasure of working with him on the Queen Charlottes, where he was as versatile as ever in his old capacity as donkey doctor. I noticed boys around Cumshewa Inlet referred to him as "Sinbad the Sailor." I found out why – it was this way:

One stormy night the tug *Northwest* was adrift in Hecate Strait with seven men aboard including Harry Todd. The skipper had radioed for help, but they would have piled up on the rocks of Cumshewa Head had Todd not come to the assistance of the chief engineer and got the engine going just in time to avert disaster.

A Queen Charlotte Islands skid road in winter. The donkey on the beach is used to skid logs down the hill. (VPL 3980)

After the engine was fixed they couldn't pick up anchor so Todd bucked off the anchor chain and gave the skipper a few pointers on navigation until they were rescued by a larger tug. Next day in camp the boys asked him how come he knew so much about the ways of ships, to which he replied, "Ah that's nothing for an old salt like me. I steered across Hecate Strait before you punks were born!"

Harry Todd at the age of sixty-seven retired from punching donkey. The old ten-gauge shotgun backfired and impaired his hearing. "Shouldn't have been using smokeless powder in it," he told me as he showed me the old relic. "They ain't like donkey boilers — they won't take the punishment. Guess I shoulda known better." When I last saw him, he was standing in his well-kept garden surrounded by beautiful flowers at his home near Victoria, where he raised prize chrysanthemums. On being asked why a high-ball donkey puncher like him had gone in for flowers, he told me: "I don't believe in waiting until I'm in the Marble Orchard to push up daisies. I'm going to enjoy 'em while I'm still alive!"

FRED GALLANT
THE BULL COOK

For a great many years the name of Fred Gallant has been known in the logging camps of the BC coast. It was on the Queen Charlotte Islands where I first made his acquaintance. He had, by then, become a member of the honourable and ancient Order of Bull Cooks. But in the good old days, back around 1910, he was a boomer brakeman who worked with Charlie Cathey and Gunner Jacobson over at Chemainus.

In those early days of railroading in the woods of Vancouver Island, the track was laid on the skid road. Then an old Climax locie dragged the logs between the rails. Fred told me he was brakeman on just such a show at Chemainus when it was the rule never to let an empty locie come down the heavy grade without a few logs in tow. Mr. Palmer, the manager, had issued strict orders this rule

Victoria Lumber & Manufacturing Company's Climax No. 3 at Camp 5, west of Ladysmith in 1907. Engineer Charlie Cathey is to left of smokestack. (UBC BC 1930/145)

Falling spruce at Allison Logging's Cumshewa operation in 1937. (Leonard Frank photos, VPL 6279 & 6280)

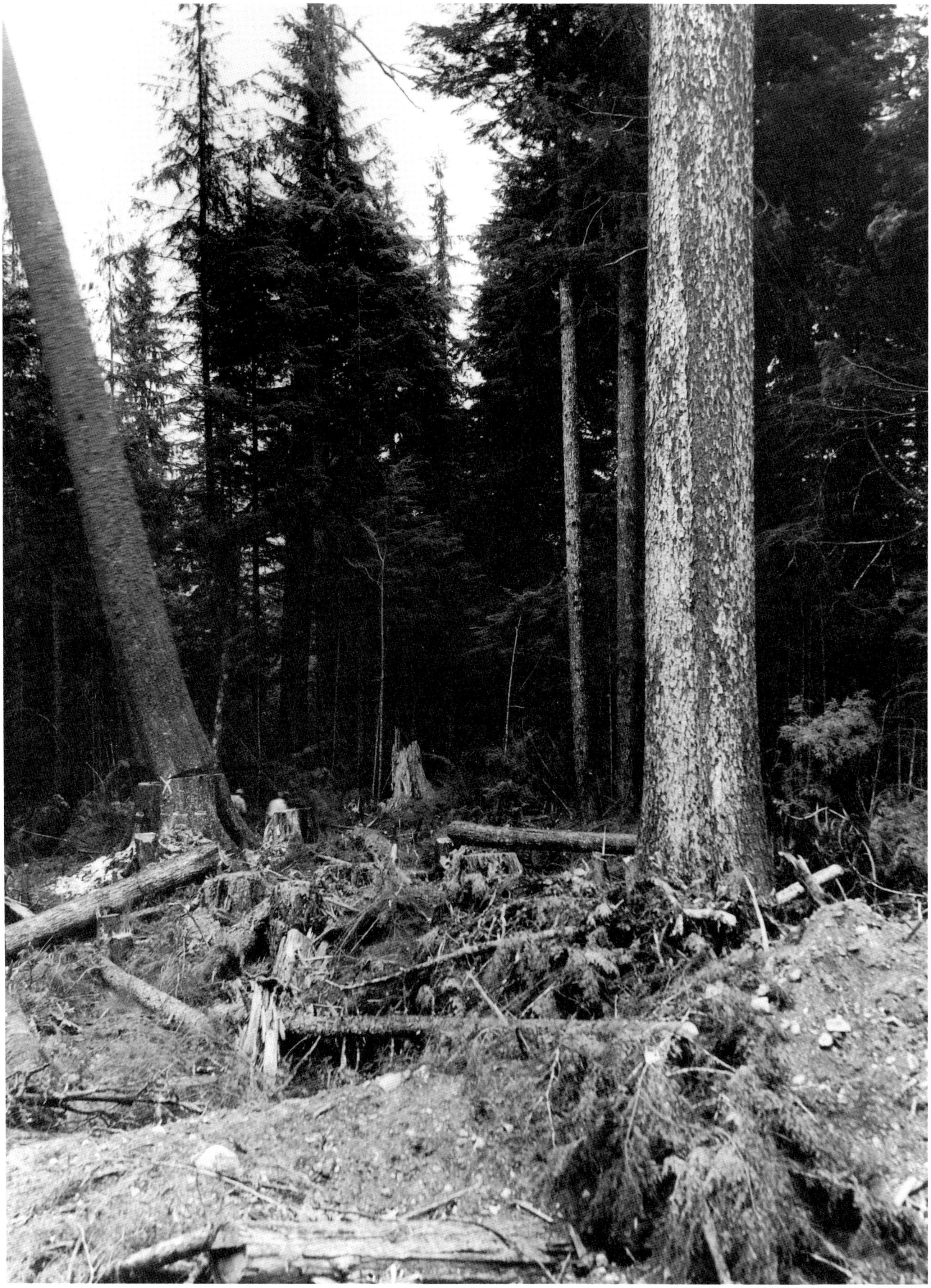

should never be broken as the locie might run away on the twelve percent grades.

One Saturday night when the locie arrived out in the woods for the last trip there were no logs on the landing. So it looked as if the crew would have to leave the locie out in the woods all night and walk back to camp. Fred, realizing there was a lot of good beer going to waste in Chemainus, decided to sand the rails. The boys did just that and made the trip back to camp without the required logs in tow. Mr. Palmer was furious but the brakeman let him rave until he cooled off a little and then said: "Look, Mister, I'm a brakeman, not a PF man. Any grade a locie can go up I can bring her down again."

They had no air brakes in those days and the disconnected trucks were coupled by link and pin. You had to take a lot of chances. Well, Fred took a chance too many one time and got a "rooster" driven through his leg which put him the hospital for several months. Braking on a modern logging train, according to Gallant, was child's play compared with the "old days."

During the first World War Fred Gallant went overseas and was taken prisoner. The Germans put him to work on a farm hoeing potatoes. One day he was brought up before the Commandant and charged with making love to a fraulein in the farmhouse. Fred was in quite a spot until he promised to confine his lovemaking to the barn and he kept out of trouble after that.

Like most of the old-timers Fred was apt to carry his drinking to excess. One time in Vancouver, when he was down and out with the DTs, he swore up and down he could see the trees on the boulevard coming at him waving their arms. Shortly after he was cured, a friend took him to see a circus. His friend noticed that Fred was fascinated by the antics of a large snake and asked him why he was so interested.

"Look, Mister," said Fred, "I've seen so many snakes lately I just thought I'd have a look at the real thing and, do you know, these are as natural looking as any I've seen."

It was at Allison's Camp on the Queen Charlottes that most of us really got to know Fred Gallant, the Bull Cook. By then he weighed 275 pounds and had a gargantuan appetite. I've seen him stow away ten eggs for breakfast and be back for a mug-up before ten o'clock.

At Allison's Fred had a pet crow and he used to talk to the crow by the hour. If he wasn't talking to his gas saw he was chatting away to the crow. It was a common sight to see Fred pushing his rubber-tired wheelbarrow with the crow perched on top of the load and Fred mumbling away to the ebony-feathered bird.

But Fred had other things on his mind besides the bird. One day he approached A.P. Allison and asked the boss to do him a favour.

"Sure, old timer," answered A.P. "What is it you want?"

"Well, it's like this," Fred mumbled. "If ever you build another camp would it be too much trouble to lay it out so as the wood yard would be up there somewhere, and the bunkhouses would all be downhill? That way I wouldn't have so much trouble wheeling the wood up to the bunkhouses. That hill is getting steeper all the time."

   And so the hill got steeper and steeper. Then, on Christmas Day 1944, old Fred took a stroke and died a few days later in Queen Charlotte Hospital. He was buried in a rocky grave at Queen Charlotte City alongside another old-timer, Boxcar Pete, and I never did find out what happened to his pet crow.

Allison Logging's Cumshewa Inlet camp, which later became the Aero Timber camp. (VPL 6272)

OLD HICKORY PALMER

The newly appointed manager of the old Chemainus mill, E.J. Palmer, was sizing up the railroad situation the first day he landed in Chemainus. He climbed aboard the old No. 21 and sat in the fireman's seat while the fireman was busy stuffing shingle hay and slabwood into the firebox of the ancient iron horse. This was railroading the like of which he'd never seen before. He didn't mind having his clothes burned off him, but it was adding insult to injury when Len Cary, the fireman, ordered him from the sea-box and told him that hoboes usually rode on the tender.

That was back in January of 1890. Palmer had just arrived from Sandpoint, Idaho, where he had been sawmilling in the short-pine

Victoria Lumber & Manufacturing Company Chemainus sawmill in 1902. (CVA Out P322B N809)

Loading lumber at the Chemainus mill in 1912. (UBC BC 1930/146)

country. But railroading had always been his big love. He was brakeman on the Chicago & Northwestern Railway in 1876. It was when he had been promoted to conductor on the westbound out of Chicago that he got to chatting with a big shot who took a fancy to him and offered him the job of managing the iron mine at Ashford. Conducting a passenger train wasn't so "cushy" since ticket-spotters had been rising the varnish, so E.J. Palmer decided right then and there to quit railroading and become a big business executive.

Back in those days a man had to be hard-boiled to be successful, and it wasn't long before E.J. Palmer became known as Old Hickory Palmer. Yet with all his success he felt as though something was lacking in his life. Maybe it was the wail of a locomotive whistle on

Victoria Lumber & Manufacturing Company's single-drum roader donkey with its line horse, west of Chemainus, 1902. (BCARS 53606)

a frosty night, or it might have been a phase in his life yet to be lived. But it wasn't until 1889 that he really found his life's work and was happy. It was then that he quit the iron business and went into the lumber business in Idaho. When the Weyerhaeuser interests took over at Chemainus, BC, Old Hickory Palmer was the man appointed to come to BC to run the big Douglas fir show over on the island.

And now to be told by a punk of a fireman that hoboes should ride the tender!

Yet, being a railroad man at heart, he was all for the kid and later promoted him to locie engineer. That was typical of Palmer. A.P. Allison, Matt Hemmingsen and many others of BC's leading lumbermen got their start in life from E.J. Palmer in much the same manner as did the locie fireman. Although he was a hard-boiled executive, Palmer never held a grudge and believed in giving credit to a man where credit was due.

Back in those early days, while many of the companies were still using bull teams and horses, Palmer introduced steam donkeys and railroad methods into the woods. It was over at Chemainus where railroad steel was laid on the skid road and Climax locies went roaring along, dragging long strings of logs between the rails.

We are told that in 1909, in both Chemainus and at Comox, the skidder system was first introduced into the BC woods. Palmer is said to have imported skidder crews from the swampy country of Louisiana. It was back in 1900 that Palmer's outfit first started logging the celebrated Cowichan Lake timber. It was Palmer's men who drove the Cowichan River to bring the first of the famous Cowichan timber to tidewater. The Cowichan River drive was never quite successful as the timber was too big, and it was Palmer who started to build a huge flume – to flume the logs from Cowichan Lake to the salt chuck. But being a railroad man he abandoned the project after negotiating with the Esquimalt & Nanaimo Railway

Weist Logging's Shay, shown near Port Alberni in 1912, was restored by Bob Swanson and is now at the Alberni Valley Museum. (VPL 5794)

The Chemainus mill in 1925. It was replaced with a modern mill in the 1980s. (UBC Humbird Papers AXV A9-3)

to build a branch line into Cowichan Lake, and they hauled logs over that branch line for about eighty-five years.

A sawmill that cut a hundred thousand feet of lumber a shift was a big mill back in those days. But it wasn't long before steam log turners, live rolls and double-cut head-rigs were introduced by the railroad-minded lumberman. By 1912 his outfit was second to none in the province and was turning out a quarter of a million feet in eight hours.

By then the original locomotive No. 21 had outlived its usefulness and Mr. Palmer had it mounted alongside the office as a monument to his former days. A good-sized alder was growing on the tender when one day Palmer decided to ship the locie — alder tree and all – to Victoria and have the locie rebuilt so he could put it to work in the woods once more. But before the ancient engine was fully renovated, the great fire of 1923 swept the big sawmill, burning it to the ground.

The big fire of November 17, 1923 seemed to break the old lumberman's heart. He didn't live to see his pet locomotive steam back into Chemainus, for he died the day after Christmas that year.

Thus ended the career of the grand old dean of BC lumber, Edmond James Palmer, who was born in Pennsylvania in 1856. His life's work was not in vain, for out of the ashes of his labour was erected at Chemainus one of the largest sawmills in the British Empire, a tribute to his memory.

HENRY NORMAN THE SKIDDER RIGGER FROM LOUISIANA

Henry Norman, the skidder rigger from Louisiana, drove his spurs into the thick flaky bark of the Douglas fir he was climbing. His topping axe dangled from his belt as he smartly worked his way up to the first limb. A few deft swings of his axe and the branch floated to the ground. He continued to climb, lopping off limbs until he paused and sank his axe into the trunk 150 feet up from the ground. Then he started in to top her.

A crowd of awestruck loggers craned their necks looking up at him until their eyes watered. When the top leaned dizzily over and plummeted to earth, the rigger leaned back in his belt as the spar swung in a great circle. Then he sat on the top and rolled a cigarette. It was not until he climbed down to the ground again and unbuckled his belt that the Comox loggers breathed easy. This was a memorable moment in the history of BC logging, for it was the first skidder tree ever to be topped in the Comox Valley, if not in the whole of BC.

This took place in 1909 up at Comox after Henry Norman had arrived from the cypress swamps of Louisiana. He had come to set up the skidder system for the Comox Logging Company. No wonder the ground-lead loggers were amazed by the antics of this high-climbing rigger with nerves of steel. The overhead system of logging was a new and revolutionary idea in the BC woods, and it was up to Henry Norman to make a success of it. That was why the Lidgerwood people had sent him to BC with the new skidder machines and rigging.

Henry was born in Finland in 1885 and was rigging on windjammers in that country when he was fifteen years old. He was in the port of New Orleans in 1903. Then he left his ship to go rigging on the skidders in the great alligator-infested swamps of Louisiana. The system had just been developed and riggers were as scarce as hen's teeth.

Down in that swampy country where the long moss hung on the cypress trees, they didn't have a slack-pulling line on the carriage. But they had a gang of workers to pull slack on the tong line, and the rigger's biggest trouble was to keep them from getting killed. They would dive under logs and into holes like gophers, and the only way to keep track of them was to count heads once in a while.

Opposite: Rigger topping a tree at Brooks, Scanlon & O'Brien's Stillwater camp in 1926. (VPL 1582)

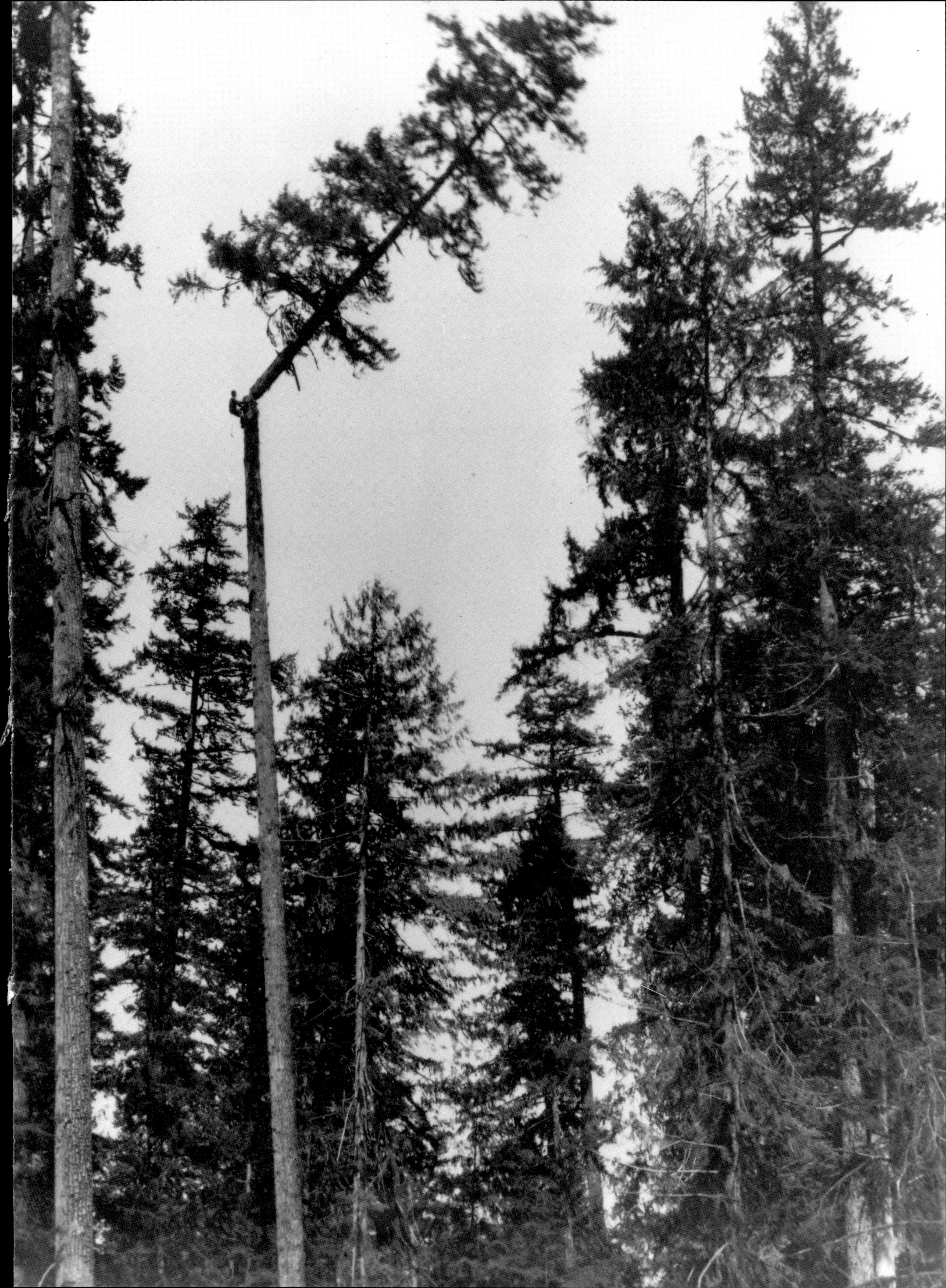

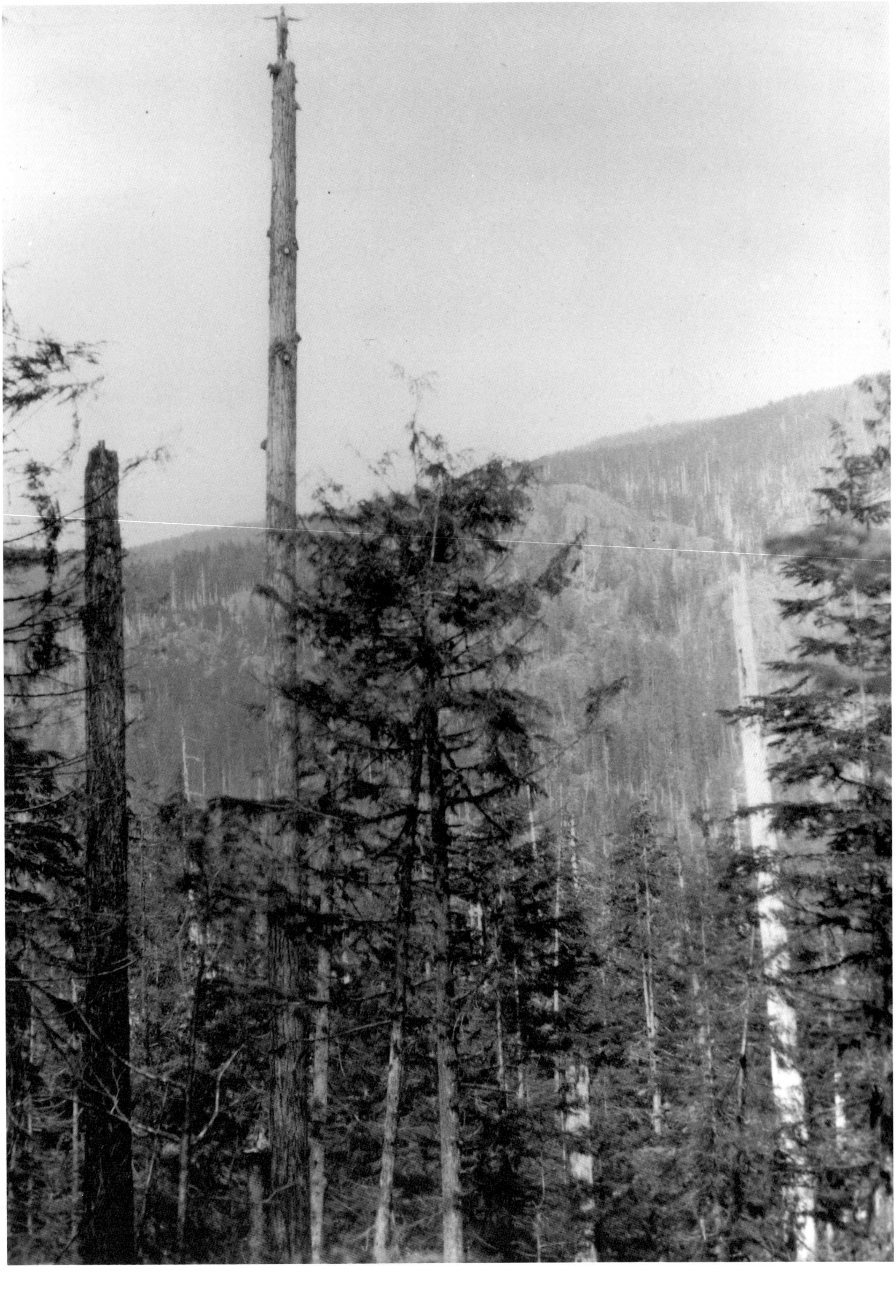

Rigger Jack Hulbert at Stillwater, 1926. (VPL 1583)

When the cable was changed to a new road, the rigger went through and chucked out the alligators. This was done by throwing the tongs on them and yarding the alligators out. They were dropped just back of the pile but the biggest trouble was to get the tongs out of them. The chaser just had to have good caulks in his boots around those outfits or he didn't last very long.

The bunkhouses had eighty men in each bunkhouse. Double-deck bunks were arranged around a big open fire in the centre. There was a hole in the roof to let out the smoke. In the morning the first fellow up got the best pair of socks and the last up got the ones with the most holes. And when a logger spoke of town down in that country he meant New Orleans. They had just finished a setting one day in 1908 when the boss brought the Lidgerwood representative up to Henry Norman. He asked Henry how he'd like to

Opposite: Having topped the spar, the rigger climbs up and poses for the photographer. (VPL 1581)

A high rigger celebrates the completion of his work at Abernethy & Lougheed's Camp 3 on Alouette Lake, 1926. (VPL 1487)

Opposite: Servicing the rigging at Bloedel, Stewart & Welch's Menzies Bay camp, 1926. (VPL 1498)

take a few new skidder machines out to the Pacific Northwest and set them up and rig them. Henry took the job and it was through this that he landed in Comox in 1908 to top the first skidder spar.

Well, the system was a great success on the coast, even in its early form without a slack-puller and with chains instead of tree straps. The early skidders didn't use a tree jack on the back spar. They wrapped the back tree with the cable and didn't bother to top it and used only one guyline. The system was used in this form for a good many years until the Berger slack-pulling skyline carriage was invented.

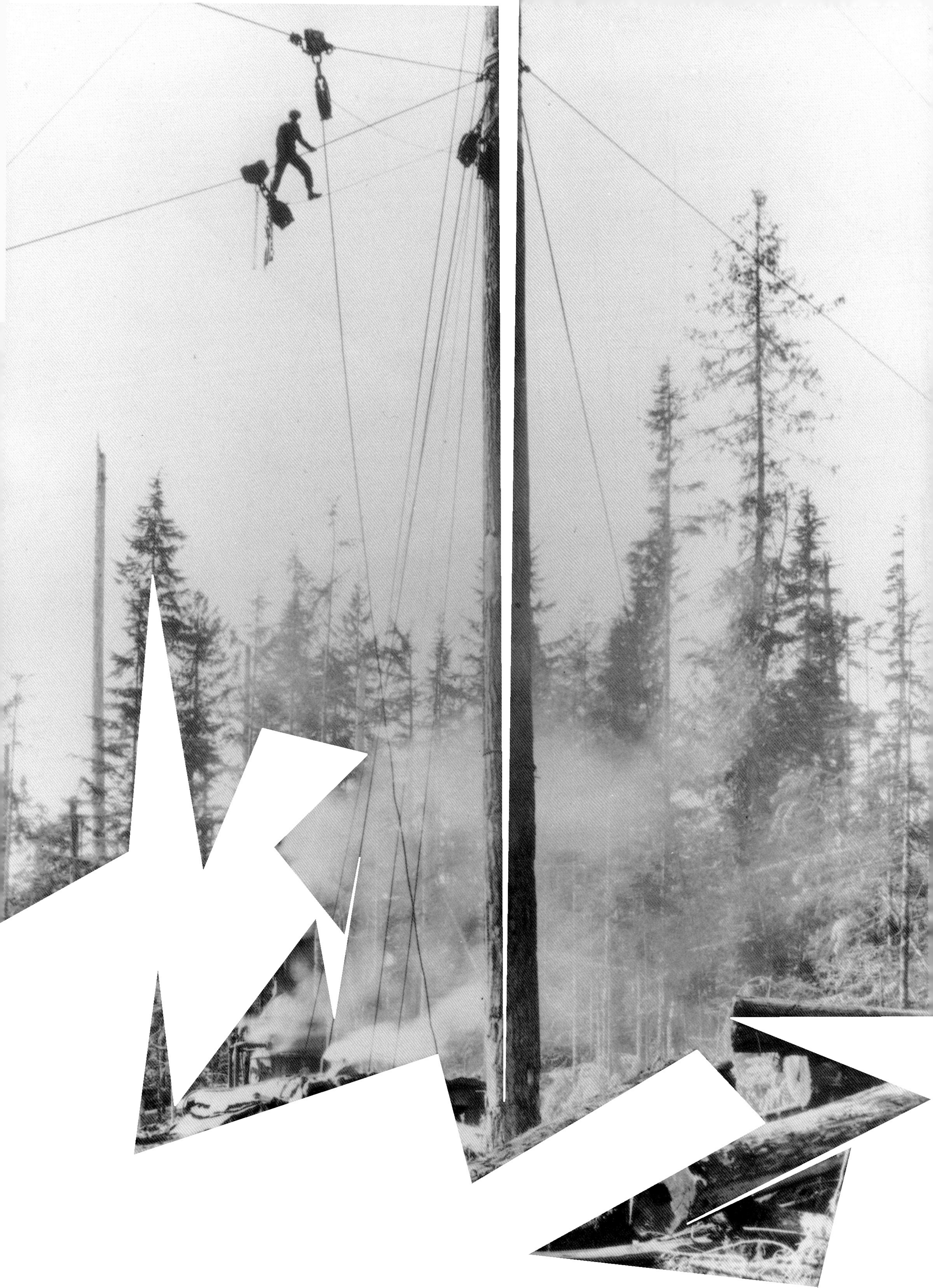

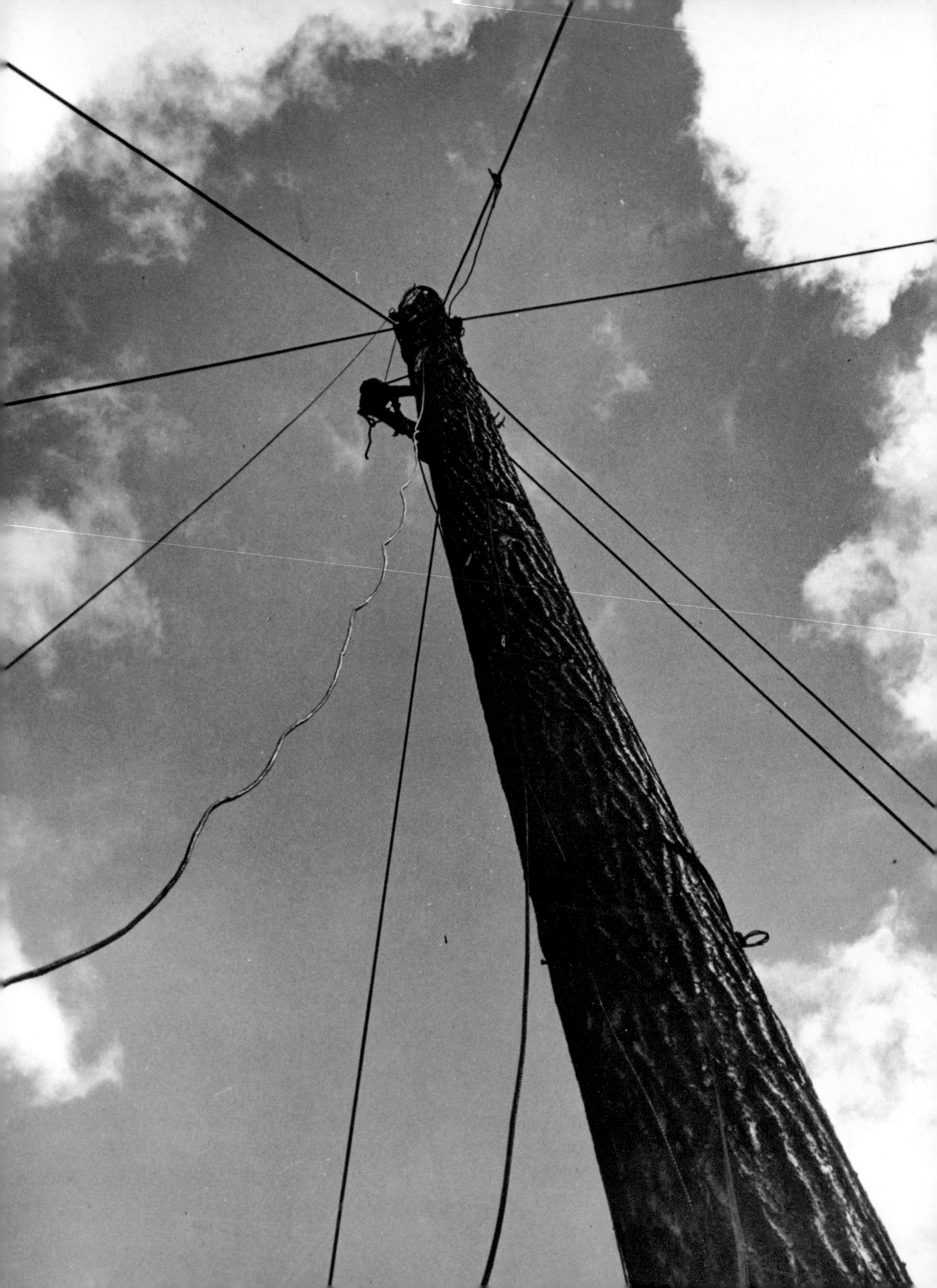

Henry Norman's next assignment was for Matt Hemmingsen at old Camp 6, back of Ladysmith, in 1912. He had worked with Matt up around Comox and when Matt had wanted to high-ball the new skidders at Camp 6 he'd sent for him. When Matt went into the Cowichan Lake country it was Henry Norman that he chose to rig up the first skidders.

The first skyline ever to be rigged was at a place called Wardroper Creek on Cowichan Lake. It was Henry who topped the tree and hung the rigging. It was Henry who did the actual rigging of the first skidder on Cowichan Lake in 1916. It was no wonder Matt Hemmingsen made a foreman out of him after that.

Jessie James hired the services of this high-climbing Finn when he purchased his skidders back in 1920. Many strange things happened at Jessie James's camp.

One Sunday Henry got under the spell of John Barleycorn and took to the woods. When he didn't come in for supper, a few of the boys went out to look for him. They found that he had taken his belt and spurs and topped a big fir. When they saw him he was standing on the top waving his arms around, but they coaxed him down with a chew of snoose.

I was over at Camp 6 on Cowichan Lake when I ran across this famous pioneer high rigger. He had just come in from the woods. He had on a rumpled bone-dry hat, a tin coat and a pair of logger's pants suspended by the customary piece of cord through the braces. There was a can of snoose in his shirt pocket and a trickle of snoose drooling from his jowls. He told me that he had gone to Russia to set up the high-lead system in 1923, but those Russians were so stupid and had been logging with horses so long that he gave them up in disgust and came home to Cowichan Lake.

And when I asked him if he figured on going home to Finland to retire, he answered: "I don't know why in heck a man 'ud go to Finland to retire when he's found out BC is the best spot on earth."

Opposite: "Hanging the jewellery" at Victoria Lumber & Manufacturing in the 1940s. (Jack Cash photo, UBC BC 1930/549/64VL 4002)

INDEX

Abernethy Lougheed, 112, 113, 120, 154
Aero Timber Camp, 143
Alberni, 38
Albion Iron Works, 81
Allison, A.P., 18, 22, 142, 146
Allison Logging, 140–41, 142, 143
Alouette Lake, 113, 120, 154
Anderson, Charles John (Charlie, the Lucky Swede), 42–46
Anderson, P.B. (Pete), 48–54

Barbrick, Bill, 72–79
BC Mills Timber & Trading Company, 34, 84
BC Provincial Police, 126–27
Bernard Timber Company, 63, 102
Bloedel, Stewart & Welch, 39, 61, 74, 76, 77, 114, 155
Boxcar Pete, 27, 143
Brodeur, Pea Soup, 59
Brooks, Scanlon & O'Brien, 62, 74, 150
Broomhandle Charlie (Snowden), 80–84
Brown, Chief Billy, 86–89
Brunette Sawmill, 43, 44
Buckley Bay, 20–21, 26, 27
Bull Sling Bill (William Strausman), 18–27

Campbell River, 95–96
Capilano Timber Company, 17, 38, 62, 92
Cary, Len, 67–71, 144
Cathels & Sorenson Logging, 107, 122, 124
Cathey, Charlie, 65, 139
Cedarville, Washington, 49
Chemainus, 67–71, 111, 118–24, 139–43, 144–49
Climax locomotive, 37, 38, 40, 83, 96, 124, 126, 147
Coburn, John, 81, 100, 104
Comox, 150

Comox Logging & Railway Company, 136
Copper Canyon, 65, 101, 116
Cowichan Lake, 16, 147–48, 95, 109–110, 115–16, 121, 157
Cumshewa, 140–41, 143
Curly, 33, 36
Curtis, Lea, 35

Dillon, Spike, 59

East Thurlow Island, 33, 35
Elk Bay, 131
English, Norman, 37
Esquimalt & Nanaimo Railway, 80–82, 116, 147–48

Fair, Al, 43
Flat Island, 50
Flying Dutchman, 126–28

Gallant, Fred, 139–43
Gambier Island, 32
Gastown, 28
Gibsons, 32
Goat Ranch, 38
gold rush, 42–45, 48
Graham, Greasy Bob, 35
Grassy Bay, 59
Great Onion River Camp, 18

Haida, 85–89
Hanney, George, 125–28
Harry Todd, 134–38
Harvie, Robert, 81
Hastings Mill, 28–36, 50
Hecate Strait, 123
Hemmingsen, Mathias (Matt), 16, 118–24, 146, 157
Hemmingsen-Cameron Timber Company, 122–24
Hillcrest Lumber, 98
Holbrook, Stewart, 50
Hosko, Charlie, 100
Hudsons, John, 32
Hulbert, Jack, 152

Humbird, 118
Hutton, Curly, 37–41

Industrial Timber Mills, 96
International Timber, 96, 97

Jacobson, Gunner, 139
James, Jessie, 108–111, 121, 157
Jemerson, John, 129
Jingle Pot Mine, 83
Johnnie-on-the-Spot, 121
Jones, Dirty Face, 131, 133
Jones Creek, 99
Jordan River, 104

Kingcome Inlet, 46
Knox Bay, 52, 53

Ladysmith Lumber Company, 81
Ladysmith, 64–66, 82, 116–17, 125, 126, 139, 157
Lake Logging, 98
Lamb, John Blacklock "Daddy," 41, 55–58
Lamb, Tom, 57–58
Lamb Lumber Company, 55–58
Lang Bay, 61
Lidgerwood skidder, 64, 65, 150–154
Long, Jack, 95
Loughborough Inlet, 15, 90, 91
Lucky Swede (Charles John Anderson), 42–46
McGarrigle, Charlie, 126
McGee, Sam, 59
McGrew, Dangerous Dan, 59
McMann, Orville, 64–66
McMillan, Smoothy, 94
MacMillan Bloedel, 70
McNair, Sandy, 33
McNeff, Jack, 129
McNeil, Jack, 125
Magedonia, 41
Massett Timber Company, 20, 26
Menzies Bay, 37, 55–58, 79, 114, 155
Meyland, Chris (the Bull of the Woods), 100–103

Milligan, Big Jack, 104–107
Milligan, Orville, 104
Milligan's Camp, 104
Moore, George, 39
Morrison, Red, 129–33
Myrtle Point, 39, 61, 74, 76, 77

New Westminster, 43, 44
Nimpkish Lake, 37
Norman, Henry, 150–57
Northwest, 137

Olar, Wilfred, 43
Oleson, (Rough House) Pete, 69, 109–110, 121
Oleson, Sam, 35
Orford Bay, 102

Palmer, E.J. (Old Hickory), 126, 141–42, 144–49
Parks, Al, 43
Peter, Charlie, 41
Peters, Bill, 50
Phelan, Hank, 97
Phelps, Jack, 15
Pioneer, 80
Port Alberni, 147
Porter locomotive, 70
Port Neville, 63
Port Renfrew, 107, 122
Powell River Company, 46, 47
Pringle, Reverend George C.F., 45, 59–63

Quatsino, 105
Queen Charlotte Islands, 18, 20–22, 24, 25–27, 76–77, 85–89, 123, 129, 131, 132, 133, 135, 137–38, 139, 140–42
Quilchena, 16

Reamy, Saul, 19, 28–36
Rock Bay, 19, 29, 34, 60, 84, 115
Rough House Pete (Oleson), 69, 109–110, 121
Royal City Planing Mills, 81

Salmon River Logging Company, 52
Schakke's Machine Works, 37
Seaford Company, 129
Seattle Red (Danny Swanson), 14, 103, 112–17
Sea Wolf, 76–77
Sechelt, 32
Shannon, Whiskey, 35
Shawnigan Lake, 103

Shay locomotive, 39–40, 67, 68, 69, 97, 100, 147
Short, Inspector Jack (Honest John), 37, 94–99
short stakers, 15–17
Six-Spot, 67–71
Skagway Jack, 20
Sky Pilot, 59, 61
Smith, Harry, 85–89
Smith, Herb, 106
Smith, Jack, 79
Smith, Sid, 79
Snowden, Broomhandle Charlie, 80–84
Stillwater, 62, 74, 150, 152
Strausman, William (Bull Sling Bill), 18–27
Stud Cat, 110
Surrey, 81
Swanson, Danny (Seattle Red), 14, 103, 112–17

Texada Island, 42, 44–45, 59
3-Spot, 37
Timberland Development Company, 70
Todd, Harry, 103
Trailing, Slim, 64–65
Truthful Thomas, 18
Tsolum River, 118–19
Tyee Lake, 136

Union Bay, 126–28
Union Steamships, 53

Vancouver, 28–29, 73
Victoria, 109
Victoria Lumber & Manufacturing Company, 40, 65, 68, 69, 77, 78, 82–83, 95, 99, 109, 111, 116, 121, 126, 127, 139, 144, 145, 156

Weist Logging, 147
Wellington, 80
West Thurlow Island, 51
Weyerhaeuser Company, 49, 146
Williams, Parker, 126
Wilson, Eight-Day, 14–17, 121

Young Iron Works, 115

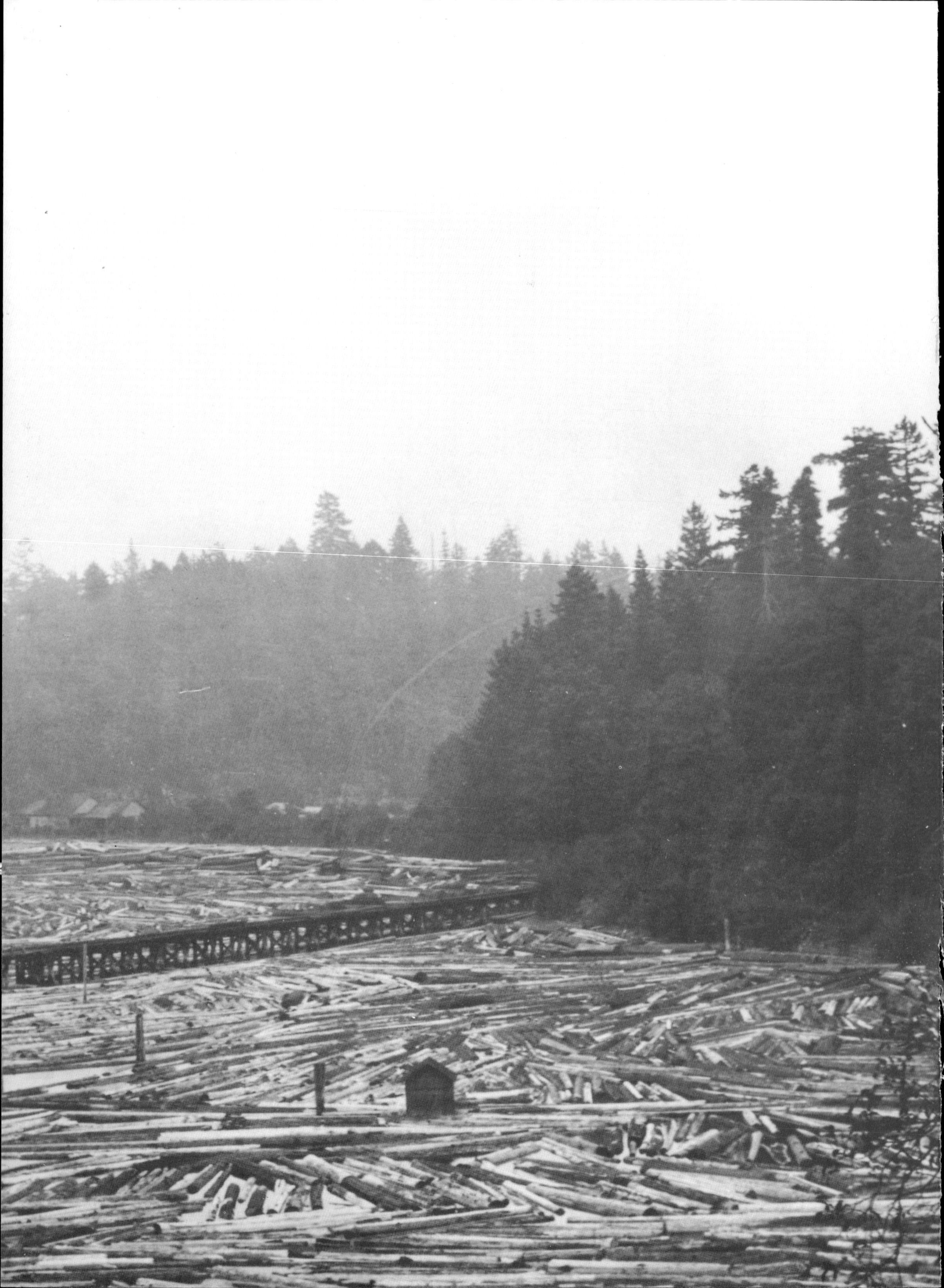